CRTD-Vol. 75

Technical Peer Review Report

Report of the Review Panel

Salt Waste Processing Facility Technology Readiness

ASME International

Institute for Regulatory Science

DISCLAIMER

This report was prepared through the collaborative efforts of the American Society of Mechanical Engineers (ASME) Center for Research and Technology Development and the Institute for Regulatory Science (referred to thereafter with the collaborators as the Society) for the Office of Science and Technology Development of the U.S. Department of Energy (referred to hereafter as the Sponsor).

Neither the Society nor the Sponsor, or others involved in the preparation or review of this report nor any of their respective employees, members, or persons acting on their behalf, make any warranty, expressed or implied, or assume any legal liability or responsibility for the accuracy, completeness, or usefulness of any information, apparatus, product, or process disclosed, or represent that its uses would not infringe privately owned rights.

Information contained in this work has been obtained by the American Society of Mechanical Engineers from sources believed to be reliable. However, neither ASME nor its authors or editors guarantee the accuracy or completeness of any information published in this work. Neither ASME nor its authors and editors shall be responsible for any errors, omissions, or damages arising out of the use of this information. The work is published with the understanding that ASME and its authors and editors are supplying information but are not attempting to render engineering or other professional services. If such engineering or professional services are required, the assistance of an appropriate professional should be sought.

Statement from By-Laws: The Society shall not be responsible for statements or opinions advanced in papers . . . or printed in its publications. (7.1.3)

ISBN No. 0-7918-0213-2

TABLE OF CONTENTS PAGE NO.

Preface

This report contains the results of a peer review performed jointly by the American Society of Mechanical Engineers (ASME) and the Institute for Regulatory Science (RSI). Based on a request from the Savannah River Office of the U.S. Department of Energy (DOE), a Review Panel (RP) was established to peer review "Salt Waste Processing Facility Technology Readiness". Consistent with the ASME/RSI process, the following RP was appointed by the Peer Review Committee for Energy and the Environment (PRCEE) of ASME:

Jimmy T. Bell, Ph.D.
Nathan H. Hurt, P.E., Chair
Donald A. Orth, Ph.D.
Francis J. Patti
Joseph R. Peterson, Ph.D.
Alfred Schneider, Ph.D.

During the period covered by this report, the ASME PRCEE overseeing the peer review consisted of the following individuals:

Charles O. Velzy, Member of Executive Panel (EP) and Chair
Ernest L. Daman, Member of EP
Nathan H. Hurt, Member of EP
A. Alan Moghissi, Member of EP; Principal Investigator of the Peer Review Program
Gary A. Benda
Erich W. Bretthauer
Irwin Feller
Robert A. Fjeld
William T. Gregory, III
Peter B. Lederman
Jeffrey A. Marqusee
Lawrence C. Mohr, Jr.
Goetz K. Oertel
Glen W. Suter, II
Cheryl A. Trottier

The supporting staff included the following individuals:

Michael Tinkelman: Director of Research at the Center for Research and Technology Development of ASME in Washington, DC; and Administrative Manager of the ASME PRCEE.

Betty R. Love: Executive Vice President, RSI, Columbia, MD; and Administrative Manager of the Peer Review Program.

M.C. Kirkland: Vice President for Southeastern Region, RSI; and Project Manager.

Sorin R. Straja: Vice President for Science and Technology, RSI; and Technical Secretary.

Sharon D. Jones: Director of Training Programs, RSI; and Manager of Review Panel Operations.

The biographical summaries of the members of the RP, the PRCEE, and the technical staff are located at the end of this report.

The Review Criteria were provided by the Project Team—consisting of principal investigators, project managers, and others involved in the project. These criteria were slightly revised by the Technical Secretary and approved by the Project Team. The RP received documents describing various aspects of the project. The summary of the project included in this report was prepared by the Technical Secretary using the same documents that had been provided to the Review Panel. The Project Team received the project summary for review and approval. In addition, the staff of RSI undertook the task of preparing a listing of acronyms. This list, as reviewed and approved by the Project Team, is also included in this report.

The RP met from October 27, 2003 through October 31, 2003, in Augusta, GA. At the beginning of the meeting, the RP and the Project Team were introduced to the ASME/RSI peer review process. The introduction was followed by presentations by the Project Team and a discussion period. The agenda of this meeting is shown in the Appendix. At the end of the meeting, the RP met in an executive session to develop its strategy for writing the report. Subsequently, the RP wrote its findings and recommendations. The *Report of the Review Panel* was then copyedited.

Consistent with the procedures established by the ASME/RSI process, the copyedited *Report of the Review Panel* was provided to DOE for identification of potential errors, misunderstandings, and areas of ambiguity. The Technical Secretary contacted the members of the RP reporting the comments received from DOE which were considered by the RP.

In addition, the Project Team chose to revise the project summary significantly. The revised summary is included in this report. Furthermore, during the closeout session, it became clear to the Project Team that relevant information—particularly that related to deposition of particulate matter in transfer pipes—had not been presented to the RP. Accordingly, the DOE provided a supplement to the project summary. This supplement is placed after the section "Findings and Recommendations" in this report.

Charles O. Velzy
A. Alan Moghissi

Appendix

ASME/RSI Peer Review
Salt Waste Processing Facility Technology Readiness

Augusta, GA - October 27-31, 2003

AGENDA

Monday, October 27, 2003

Radisson Riverfront Hotel, Augusta
Lamar C Room

8:00 a.m.	Welcome	Terrel Spears, U.S. DOE - SRS
8:10 a.m.	ASME/RSI Peer Review Process	A. Alan Moghissi, ASME/RSI
9:00 a.m.	Salt Processing Project Overview • HLW Overview • Overall Salt Processing Strategy • SWPF Project Requirements • Current Project Status	Terrel Spears
9:30 a.m.	Background on SWPF Technology Selection • Technology Risks • Down Selection Process • Risk Assessment Summary • Previous NRC Reviews and Status of Recommendations	Harry Harmon, PNNL
10:00 a.m.	Break	
10:15 a.m.	Actinide/Sr Removal	
	• Overview, Process Description, and HLW Interfaces	Sam Fink, WSRC
	• Actinide and Strontium Removal Chemistry - Monosodium Titanate - Alternative Sorbents	David Hobbs, SRTC
	• Solid/Liquid Separation - Cross-flow Filtration - Alternative Solid/Liquid Separation	Michael Poirier, WSRC
	• Analytical Monitoring - On-line Neutron Monitor - Analytical Methods	Sam Fink
	• Summary of Residual Technical Risks	Sam Fink
12:00 noon	Lunch	

ASME/RSI Peer Review
Salt Waste Processing Facility Technology Readiness

Augusta, GA - October 27-31, 2003

AGENDA

Monday continued

12:45 p.m.	Caustic-Side Solvent Extraction	
	• Overview, Process Description, and HLW Interfaces	Major Thompson, SRTC
	• Solvent Components Development and Commercialization	Peter Bonnesen, ORNL
	• Solvent Stability - Chemical and Thermal Stability - Radiolytic Stability	Peter Bonnesen
	• Simulant Flowsheet Tests	Ralph Leonard, ANL
	• Actual Waste Flowsheet Tests	Bob Pierce, SRTC
	• Centrifugal Contactor Equipment Tests	Joe Birdwell, ORNL
	• Solvent Washing	Joe Birdwell
	• Entrained Solvent Recovery	Ralph Leonard
	• Summary of Residual Technical Risks	Major Thompson
2:30 p.m.	Break	
2:45 p.m.	FW/BNFL Presentation *(restricted access)*	
	• Assessment of Technology Status and Risks	Chris Phillips, BNFL
	• Planned Technology Development	Mike Washer, BNFL
	• Questions from Review Committee	
4:00 p.m.	Break	
4:15 p.m.	Parsons Presentation *(restricted access)*	
	• Assessment of Technology Status and Risks	Jack Kasper, Parsons
	• Planned Technology Development	Alan Stephens, Parsons
	• Questions from Review Committee	
5:30 p.m.	Adjourn	

Tuesday, October 28, 2003

Radisson Riverfront Hotel, Augusta
Lamar C Room

8:00 a.m.	Executive Session
9:30 a.m.	Discussion

ASME/RSI Peer Review
Salt Waste Processing Facility Technology Readiness

Augusta, GA - October 27-31, 2003

AGENDA

Tuesday continued

10:30 a.m.	Break
10:45 a.m.	Discussion (continued)
12:00 noon	Lunch
1:00 p.m.	Executive Session
5:00 p.m.	Adjournment

Wednesday, October 29, 2003

Radisson Riverfront Hotel, Augusta
Cumming Room

8:00 a.m.	Executive Session
5:00 p.m.	Adjournment

Thursday, October 30, 2003

Radisson Riverfront Hotel, Augusta
Cumming Room

8:00 a.m.	Executive Session
5:00 p.m.	Adjournment

Friday, October 31, 2003

Radisson Riverfront Hotel, Augusta
Cumming Room

8:00 a.m.	Closeout Session
10:00 a.m.	Adjournment

Executive Summary

The U.S. Department of Energy (DOE) selected Caustic Side Solvent Extraction (CSSX) for cesium removal and monosodium titanate (MST) strike with filtration for removal of actinides and strontium as the technologies to be implemented in the planned Salt Waste Processing Facility (SWPF). Following the technology selection in 2001, extensive technology development activities for CSSX and MST/Filtration technologies continued at several DOE laboratories and universities in an effort to mature those technologies and to optimize them for use in the SWPF. In 2002, DOE selected two Engineering, Procurement, and Construction (EPC) contractors to develop competitive conceptual designs for the SWPF based on CSSX and MST/Filtration technologies. In January 2004, DOE plans to complete SWPF conceptual designs and request project Critical Decision One (CD-1, Approval of Alternative Selection and Cost Range). Beyond CD-1, a single EPC contractor will complete design, construction, and commissioning of the SWPF.

The CSSX is designed to process waste streams that have been treated to remove actinides and strontium. These radionuclides are removed at the "front end" of CSSX by batch contact of the waste solution with MST. Incoming salt solution from storage tanks containing entrained sludge solids is pretreated with MST to adsorb strontium and plutonium. The resulting slurry is filtered using a cross-flow filter, and the MST and sludge solids are sent to the Defense Waste Processing Facility (DWPF) for vitrification.

The CSSX uses a multicomponent solvent that will extract cesium ions from the caustic solution. Following cesium extraction, the solvent is scrubbed with dilute caustic to remove other salts from the solvent stream. The solvent is then contacted with a dilute acid stream to transfer cesium to the strip solution. This solution is suitable for transfer to the DWPF. The stripped solvent is scrubbed to remove degradation products prior to recycling.

The CSSX technology (for cesium removal) and monosodium titanate MST/Filtration technology (for removal of actinides and strontium) have reached the necessary technical maturity required for preliminary design for deployment at the Savannah River Site (SRS).

The Project Team has identified the residual technical risks associated with implementing the CSSX technology and concluded that the identified risks are sufficiently low to proceed with the preliminary design of the Salt Waste Processing Facility (SWPF). The Project Team has adequately defined the operating conditions and limitations of the different unit operations involved with this technology. This conclusion was reached by considering the following: 1) Chemical and Physical Properties of the Solvent; 2) Process Equipment; 3) Process Chemistry and Criticality Concerns; and 4) Batch and Integrated Tests.

The Project Team has identified the residual technical risks (plutonium sorption and removal) in MST/Filtration technology. The Project Team also adequately assessed the compatibility of the MST/Filtration product stream with the CSSX process.

The Project Team has adequately assessed the compatibility of the SWPF product streams (MST/Sludge, CSSX cesium strip effluent, and decontaminated salt solution) with the DWPF and Saltstone Facility. Specifically, the MST/Sludge can be fed to the DWPF without further treatment, and the properties of the resulting waste glass will not be affected significantly. The CSSX cesium strip effluent is sufficiently concentrated to be fed to the DWPF without additional treatment. The decontaminated salt solution can be transferred to the Saltstone Facility without further treatment.

The Project Team has adequately planned technology development activities that are anticipated to mature or optimize the CSSX and MST/Filtration technology during the design or construction stages. The proposed plans show a thorough endeavor to improve the process during Phase II, consistent with the developments during Phase I.

The Project Team has adequately addressed the recommendations provided by the two independent National Research Council peer reviews.

The Review Panel (RP) was impressed by the thorough planning; quality of experimental work; risk assessments; and engineering analyses performed by the Project Team with the participation and cooperation of several organizations; and on a tight schedule. It was evident that the Project Team has maintained a sharp focus on the principal objectives. DOE management is commended for having expanded the Project Team in Phase I through the inclusion of two competing EPC vendors. Based on a careful assessment of the information presented to the RP and the findings developed in response to the review criteria, the RP provides the following recommendations:

1. The Project Team is urged to proceed with the implementation of its plans. In addition, the Project Team should also consider Recommendations 2 and 3.
2. The Project Team should consider additional testing for CSSX using real waste for extended periods of time, preferentially coupled with contactor operation.
3. The Project Team should investigate the attrition of the loaded MST during the slurry transfer to the DWPF to assess safety concerns.

Peer Review Process

INTRODUCTION

There is consensus within the technical community on the definition, process, and key criteria for the acceptability of peer review. Peer review consists of a critical evaluation of a topic by individuals who—by virtue of their education, experience, and acquired knowledge—are qualified to be peers of an investigator engaged in a study. A peer is an individual who is able to perform the project, or the segment of the project that is being reviewed, with little or no additional training or learning.

Recognizing that peer review constitutes the core of acceptability of scientific and engineering information, virtually all professional societies of scientists and engineers have instituted formal procedures for peer review for their activities. The American Society of Mechanical Engineers (ASME), also known as ASME International, has over a century of experience in peer review. Consistent with its mission and tradition, ASME, in cooperation with the Institute for Regulatory Science (RSI), has established a peer review program devoted to the review of activities of various government agencies (ASME 2003, RSI 2003). The reports of the peer reviews resulting from this program have been published (ASME/RSI 1997, 1998, 1999, 2000, 2001a, 2001b, 2001c, 2002a, 2002b, 2002c, 2002d, 2003a, 2003b).

PEER REVIEW PROCESS

The structure of the peer review process established by the ASME/RSI team consists of a tiered system. For each specific area, the entire process is overseen by a committee. The review of specific topics is performed by Review Panels (RPs).

Peer Review Committee for Energy and the Environment

The Peer Review Committee for Energy and the Environment (PRCEE) is a standing committee of ASME formed to oversee peer review for one particular program in an agency. Its members are chosen on the basis of their education, experience, peer recognition, and contribution to their respective areas of competency. An attempt is made to ensure that all needed technical competencies and diversity of technical views are represented in the PRCEE. The members of the PRCEE must be approved by the Board on Research and Technology Development of the Council on Engineering of ASME. The PRCEE includes an Executive Panel (EP) that is responsible for the day-to-day operations of the PRCEE. Except for the EP, membership in ASME is not required for appointment to the PRCEE. As the overseer of the entire peer review process, the PRCEE enforces all relevant ASME policies, including compliance with professional and ethical requirements. A key function of the PRCEE is the approval of the appointment of members of RPs for a specific project.

Review Panels

The review of a project, a document, a technology, or a program is performed by an RP consisting of a small group of highly-knowledgeable individuals. Upon the completion of their task, the RPs are disbanded. The selection of reviewers is based on the competencies required for the specific review assignment. The number of individuals in an RP depends upon the complexity of the subject to be reviewed. The selection of a reviewer is based on the totality of that individual's qualifications. However, there are several generally-recognized and fundamental criteria for assessing qualifications of a reviewer. These are as follows:

1. **Education:** A minimum of a B.S. degree, preferably an advanced degree in an engineering or scientific field, is required for any peer reviewer.

2. **Experience:** In addition to education, the reviewer must have significant experience in the area that is being reviewed.

3. **Peer recognition:** Election to an office of a professional society, serving on technical committees of scholarly organizations, and similar activities are considered to be a demonstration of peer recognition.

4. **Contributions to the profession:** Contributions to the profession may be demonstrated by publications in peer-reviewed journals. In addition, patents, presentations at meetings where the papers were

peer-reviewed, and similar activities are considered to be contributions to the profession.

5. **Conflict of Interest:** One of the most complex and contested issues in peer review is a set of subjects collectively called conflict of interest. The ideal reviewer is an individual who is intimately familiar with the subject and yet has no monetary interest in it. Despite this apparent difficulty, the ASME and similar organizations have successfully performed peer review without having a real or apparent conflict of interest. The guiding principle for conflict of interest is as follows: *Those who have a stake in the outcome of the review may not act as a reviewer or participate in the selection of reviewers.*

Due to the multidisciplinary nature of many projects reviewed by the ASME/RSI team, rapid identification of qualified peer reviewers and their availability to participate in the review process are key ingredients for a successful program. The process used for the identification of reviewers is multifaceted. The Administrative Manager of the Peer Review Program receives recommendations from sources within ASME; previous members of the RP; sister societies; other organizations and individuals; the U.S. Department of Energy (DOE); DOE contractors; and others. However, the selection of peer reviewers is based entirely on criteria identified by ASME. The details of various aspects of peer review, including conflict of interest, can be found in the ASME *Manual for Peer Review* (ASME 2003) and the associated procedures (RSI 2003).

COOPERATION WITH OTHER PROFESSIONAL SOCIETIES

The ASME is a large professional engineering society having in excess of 125,000 members. Although the predominant discipline of the members is mechanical engineering, there are members who—by virtue of their education, training, or experience—are competent in other disciplines. The Council on Engineering includes divisions ranging from classical mechanical engineering (design, heat transfer, and power) to solar engineering; environmental engineering; and safety and risk analysis.

Despite the diverse competency within ASME, it is recognized that on occasion it will become necessary to peer review activities which include disciplines that are outside the areas of competency of ASME and its members. These disciplines may include geology, hydrology, toxicology, and ecology. Consequently, ASME has reached formal and informal agreements with its sister societies to identify qualified reviewers in areas outside of those covered by the membership of ASME.

PERFORMING ORGANIZATIONS

The Center for Research and Technology Development of ASME manages a number of scientific and engineering activities, including peer reviews. Because of ASME's conscious effort to maintain a small in-house staff, it relies upon other organizations to provide detailed project management services in its research, development, and similar activities. Accordingly, ASME and RSI joined forces in a collaborative effort to perform the peer review for the U.S. Department of Energy. While the ASME staff in Washington, DC provides the staff support for the PRCEE, the detailed management and staff support for the RPs is provided by RSI.

American Society of Mechanical Engineers

As one of the largest professional engineering societies, ASME has a long and distinguished history. Its activities are carried out primarily by members who volunteer their time in support of engineering and scientific advancement. For obvious reasons, ASME also has a paid staff to manage the day-to-day operations of such a large professional society. ASME has a detailed structure for its operation, consisting of councils, boards, divisions, and committees. The Council on Engineering has 38 divisions, including: Environmental Engineering; Solid Waste Processing; Nuclear Engineering; Safety Engineering; and Risk Analysis. The Council on Codes and Standards develops ASME codes and standards that are the backbone of many industries—including power production—worldwide. The Council on Codes and Standards is also responsible for the development of standards for activities such as certification of incinerator operators. The ASME was a founding member of the American Association of Engineering Societies and a founding member of the American National Standards Institute.

Institute for Regulatory Science

RSI is a not-for-profit organization chartered under section 501(c)3 of the Internal Revenue Service. It is dedicated to the idea that societal decisions must be based on the best available scientific and engineering information. According to the RSI mission statement, peer review is the foundation of the best available scientific and engineering information. Consequently, RSI has promoted peer review within government and industry as the single most important measure of reliability of scientific and engineering information. In its activities, RSI seeks the cooperation of scholarly organizations. Historically, a large number of RSI activities have been performed in cooperation with ASME. RSI is located in the Washington, DC metropolitan area.

Project Summary

INTRODUCTION

Savannah River Site

The Savannah River Site (SRS) was established in 1950 by the then Atomic Energy Commission, a predecessor agency of the U.S. Department of Energy (DOE). The SRS has been involved primarily in the production of tritium and plutonium in support of the nuclear stockpile. It occupies approximately 310 square miles (803 km^2), and is located in the south-eastern coastal area of the United States in the state of South Carolina (Fig. 1). It is bordered to the west by the Savannah River and Georgia, and is close to several cities including Augusta and Savannah (Georgia); Columbia; Greenville; and Charleston (South Carolina). The SRS celebrated its 50th Anniversary on November 28, 2000. Table 1 contains historical highlights at SRS. Currently, SRS is a key facility of the DOE. It is managed by the Savannah River Operations Office of DOE (SROO 2002), and operated by a consortium of contractors led by Westinghouse Savannah River Company (WSRC).

In achieving DOE's goals, SRS focuses on three mission areas as follows:

1. **Nuclear Weapons Stockpile Stewardship:** Responsible for the careful use of funds, time, talents, and other resources, especially with respect to the principles and needs of a community. Nuclear Weapons Stockpile Stewardship emphasizes science-based maintenance of the nuclear weapons stockpile. SRS supports the stockpile by ensuring the safe and reliable recycle, delivery, and management of tritium resources by contributing to the stockpile surveillance program; and by its ability to assist in the development of alternatives for large-scale pit production capability (if required).

2. **Nuclear Materials Stewardship:** Management of excess nuclear materials, including transportation, stabilization, storage, and disposition to support nuclear non-proliferation initiatives. Primary nuclear materials in this program include components from dismantled weapons; residues from weapons-processing activities; spent nuclear fuel; and other legacy materials.

3. **Environmental Stewardship:** Management, treatment, and disposal of radioactive and hazardous wastes resulting from past, present, and future operations. This stewardship includes pollution prevention and restoration of the environment impacted by site operations. Environmental Stewardship also encompasses the site's extensive natural and cultural resources.

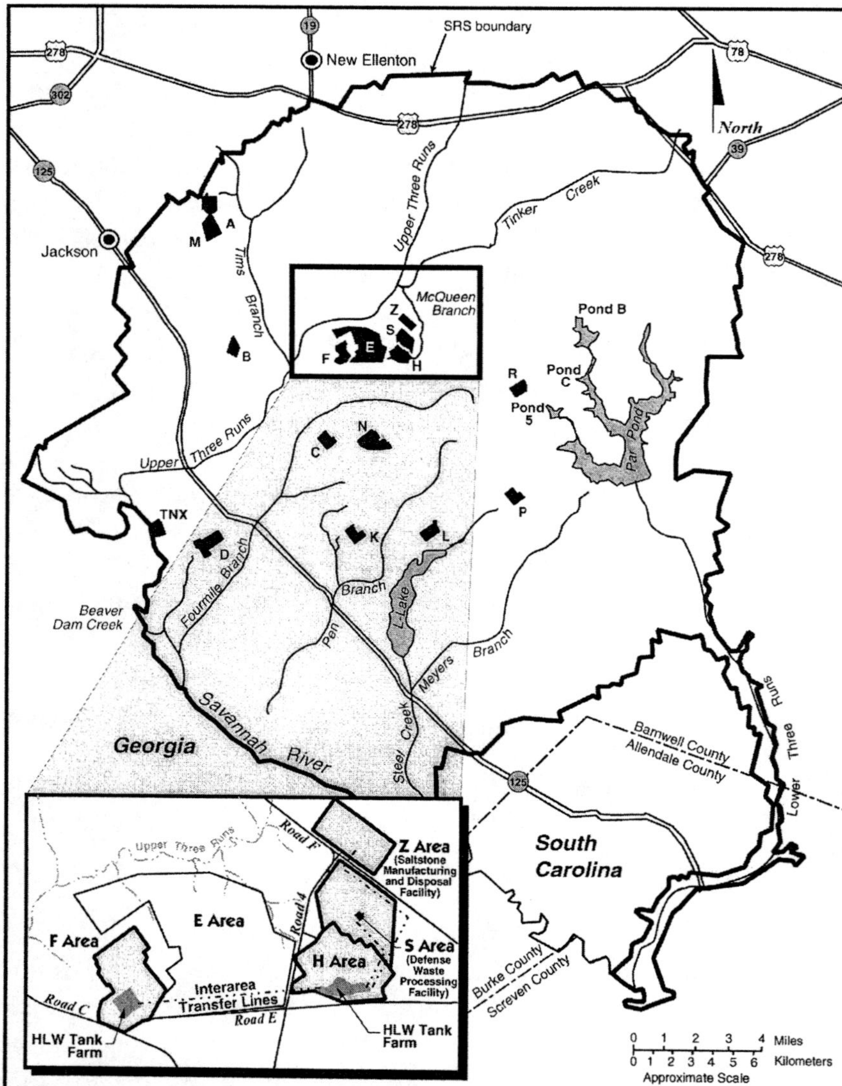

Fig. 1. Savannah River Site Map with F, H, S, and Z Areas Highlighted (DOE 2001).

Table 1. Historical Highlights at Savannah River Site (SRS 2002).

Year	Highlight
1950	E.I. duPont de Nemours and Company is asked by the Atomic Energy Commission to design, construct, and manage the Savannah River Plant. An area for the site is chosen.
1951	Savannah River Ecology Laboratory begins ecological studies of SRS plants and animals. Construction begins at the site.
1952	Production of heavy water for site reactors begins in Heavy Water Rework Facility.
1953	R-Reactor, the first production reactor, goes critical.
1954	P-Reactor, L-Reactor, and K-Reactor go critical. The first irradiated fuel is discharged. F-Canyon, a chemical separation facility, begins radioactive operations.
1955	C-Reactor goes critical. The first plutonium shipment leaves the site. H-Canyon, a chemical separation facility, begins radioactive operations.
1956	Construction of the basic plant is complete.
1963	Receiving Basin for Off-Site Fuels receives first shipment of off-site spent nuclear fuel.
1964	R-Reactor is shut down.
1968	L-Reactor is shut down for upgrades.
1971	K-Reactor becomes the first production reactor automatically controlled by computer.
1972	The site is designated as the first National Environmental Research Park.
1981	SRS begins environmental cleanup program. M-Area Settling Basin cleanup begins under the Resource Conservation and Recovery Act.
1982	Heavy Water Rework Facility is closed.
1983	Ground is broken for construction of the Defense Waste Processing Facility. Wackenhut Services Incorporated begins providing security support services at SRS.
1985	HB-Line begins producing plutonium-238 for NASA's deep-space exploration program. L-Reactor restarted and C-Reactor are shut down. A full-scale groundwater remediation system is constructed in M-Area.
1986	Construction of Saltstone begins. Construction of the Replacement Tritium Facility begins.
1987	DuPont notifies DOE that it will not continue to operate and manage the site.
1988	K-, L-, and P-Reactors are shut down. Effluent Treatment Facility begins operations to treat low-level radioactive wastewater from F- and H- Area Separations facilities.
1989	The site is officially included on the National Priority List and becomes regulated by the U.S. Environmental Protection Agency (EPA). Westinghouse Savannah River Company assumes management and operation of site facilities.
1990	Construction of a cooling tower for K-Reactor begins. Saltstone begins operations.
1991	Mixed Waste Management Facility is the first SRS facility closed and certified under the provisions of the Resource Conservation and Recovery Act. L-Reactor shut down. M-Area Settling Basin closure completed. Cold War ended and production of nuclear materials for weapons use at SRS stopped.
1992	K-reactor is operated briefly for last time and connected to cooling tower. Secretary of Energy announces phaseout of all uranium processing. Non-radioactive operations begins at the Replacement Tritium Facility.
1993	K-Reactor placed in cold-standby condition as Nation's tritium source. Non-radioactive test runs of the Defense Waste Processing Facility begin. Construction begins on Consolidated Incineration Facility. Tritium introduced into the Replacement Tritium Facility and radioactive operations begin. Workforce Transition and Community Assistance begins at SRS.
1994	SRS Citizens Advisory Board is established.
1996	The Defense Waste Processing Facility introduces radioactive material into the vitrification process. K-Reactor placed in shutdown condition. F-Canyon restarts and begins stabilizing nuclear materials at SRS.

Table 1. (cont'd).

Year	Highlight
1997	First high-level radioactive waste tanks closed.
1998	SRS is selected as the preferred site of three new plutonium missions.
2000	K Reactor building is converted to K Area Materials Storage Facility. Record of Decision is announced, selecting SRS as the site of new plutonium missions: MOX Fuel Fabrication Facility; Pit Disassembly and Conversion Facility; Plutonium Immobilization Facility. Westinghouse Savannah River Company (WSRC) Team earns the U.S. Department of Energy's top safety performance and program honor: Star Status. Defense Waste Processing Facility pours its 1,000th canister of glassified radioactive waste.
2001	Safety milestone: WSRC Operations employees worked ten-million hours without an injury resulting in time away from work. First shipment of transuranic waste is shipped to DOE's Waste Isolation Pilot Project in New Mexico. Environmental cleanup mark surpassed with the removal of industrial solvents from over four-billion gallons of groundwater. The Defense Waste Processing Facility takes the lead worldwide with the pouring of four million pounds of environmentally acceptable glassified waste. Over $44 million is saved in 2001 through the SRS employee suggestion program. SRS employee responses to the 2001 National Safety Council's safety perception survey say the site's safety culture is one of the best in the nation.
2002	After 50 years of service to the nation, the F Canyon and FB Line facilities at SRS complete their last production run to process legacy materials.

Successful accomplishment of the three stewardship missions is managed through five fundamental principles or Focus Areas:

1. Safety and Security
2. Technical Capability and Performance
3. Community, State, and Regulator Relationships
4. Cost-Effectiveness
5. Corporate Perspective

These Focus Areas crosscut all site functions, defining how work is done at SRS, with an emphasis on achieving results and continually improving performance.

High-Level Waste System

The Savannah River Site High-Level Waste System is a set of different interconnected processes operated by WSRC (Closure Business Unit). These processes function as one large treatment plant that receives, stores, and treats high-level radioactive waste (HLW) at SRS and converts these wastes into forms suitable for final disposal.

These processes currently include:

1. HLW Storage and Evaporation (F and H Area Tank Farms)
2. Salt Processing
3. Sludge Processing (Extended Sludge Processing [ESP] Facility)
4. Vitrification (Defense Waste Processing Facility [DWPF])
5. Wastewater Treatment (Effluent Treatment Facility [ETF])
6. Solidification and Disposal (Saltstone Production Facility [SPF] and Saltstone Disposal Facility [SDF])

The F and H Area Tank Farms, ESP Facility, DWPF, ETF, SPF, and SDF are all operational. The In-Tank Precipitation (ITP) Facility operations are limited to safe storage of materials. The Late Waste Facility has been tested and is in preparation for start-up of the Actinide Removal Process (ARP).

The mission of the SRS HLW System is to receive and store HLW in a safe and environmentally sound manner and to convert these wastes into forms suitable for final disposal. The planned disposal forms are:

1. Borosilicate glass to be sent to a Federal repository
2. Saltstone to be disposed on-site
3. Treated wastewater to be released to the environment

Also, the storage tanks and facilities used to process the HLW must be left in a state such that they can be closed and decommissioned in a cost-effective manner, and in accordance with appropriate regulations and regulatory agreements.

All HLW in storage at SRS must follow hazardous waste requirements. Consequently, DOE has entered into a compliance agreement with the U.S. Environmental Protection Agency (EPA) and the South Carolina Department of Health and Environmental Control (SCDHEC). This compliance agreement is implemented through the Site Treatment Plan, which requires processing of all the HLW at SRS according to a schedule negotiated between the parties.

Figure 2 schematically illustrates the routine flow of wastes through the SRS HLW System. The various internal and external processes are shown in rectangles. The numbered streams identified in italics are the interface streams between the various processes.

Incoming HLW (Stream 1) is received into HLW Storage and Evaporation facilities (F and H Area Tank Farms). The function of HLW Storage and Evaporation is to safely concentrate and store these wastes until downstream processes are available for further processing. The decontaminated liquid from the evaporators (Stream 13) is sent to ETF.

The insoluble sludges that settle to the bottom of waste receipt tanks in HLW Storage and Evaporation (Stream 2) are slurried and sent to ESP. All sludges are washed with water to reduce their soluble salt content. The spent washwater from this process (Stream 3) is sent back to HLW Storage and Evaporation. The washed sludge (Stream 4) is sent to DWPF for feed pretreatment and vitrification.

Saltcake is redissolved using techniques including slurrying, density gradient, etc. The dissolved salt solutions from this operation, and other salt solutions from HLW Storage and Evaporation (Stream 5) in the Actinide Removal Process (ARP) and the future Salt Waste Processing Facility (SWPF) are processed to remove radionuclides (actinides, strontium, and cesium). These concentrated radionuclides are then prepared for transfer to DWPF. Strontium and actinides are removed by sorption with monosodium titanate (MST), and the slurry is filtered to remove MST and entrained sludge solids. The MST and sludge solids are transferred to DWPF (Stream 8). For the Caustic-Side Solvent Extraction (CSSX) process, cesium contained in the organic phase (solvent) is stripped to an aqueous phase for transfer to DWPF, and the solvent is recycled. The decontaminated salt solution (DSS) is sent to SPF for disposal.

The washed sludge from ESP (Stream 4) is chemically adjusted in the DWPF to prepare the sludge for feed to the glass melter. As part of this process, mercury is removed, purified, and sent to mercury receivers (Stream 12).

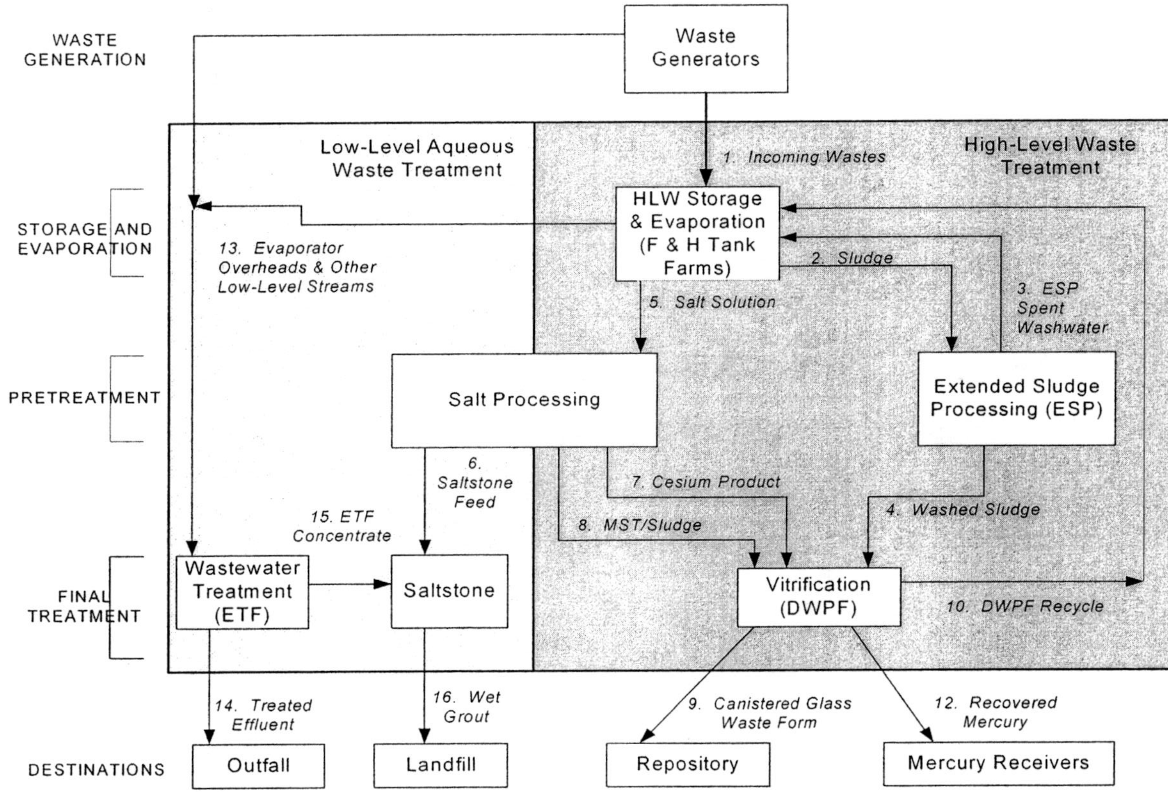

Fig. 2. High-Level Waste System Major Interfaces (Harmon et al. 2001a).

The aqueous cesium product from the SWPF is added to the chemically-adjusted sludge. The mixture is then combined with glass frit and sent to the glass melter. The glass melter drives off the water and melts the wastes into a borosilicate glass matrix, which is poured into a stainless-steel canister. The canistered glass waste form (Stream 9) is sent to on-site interim storage, and will eventually be disposed in a Federal repository.

The water vapor driven off the melter is condensed and combined with other aqueous streams generated throughout the DWPF. The combined aqueous stream is recycled (Stream 10) and transferred to HLW Storage and Evaporation for processing.

Overheads from the HLW Storage and Evaporation evaporators are combined with overheads from evaporators in the F and H area separation processes and other low-level streams from various waste generators. This mixture of low-level waste (Stream 13) is sent to the Effluent Treatment Facility (ETF).

In the ETF, low-level waste (LLW) is decontaminated by a series of cleaning processes. The decontaminated water effluent (Stream 14) is sent to the H-Area outfall and eventually flows to local creeks and the Savannah River. The contaminants removed from the water are concentrated (Stream 15) and sent to the SPF. In the SPF, the liquid waste (Streams 6 and 15) is combined with cement formers and pumped as a wet grout (Stream 16) to a vault located in the Saltstone Disposal Facility (SDF). In the vault, the cement formers hydrate and cure, forming a saltstone monolith. The SDF will eventually be closed as a landfill.

Salt Waste Processing Background

The HLW has been stored onsite in large underground tanks. The DWPF has been operating since 1996 to vitrify some of the HLW components. At SRS, HLW was generated as an acidic solution. Subsequently, it was chemically converted to an alkaline solution for storage. In its alkaline form it consists of two components: salt and insoluble sludge. Radionuclides found in the sludge include fission products (such as ^{90}Sr) and long-lived actinides (plutonium). Radionuclides found in the salt component include isotopes of cesium and technetium, as well as some strontium and actinides.

Dewatering the salt solution by evaporation, which conserves tank space, converts the salt solution into a solid saltcake and a concentrated salt supernatant. As a first step to process the salt components, solid saltcake must be dissolved by adding water, and combined with salt supernatant, to form a salt solution. An important part of the DWPF system, as designed, was to separate the highly radioactive constituents from the salt solution. The high-activity fraction removed from the salt solution would be vitrified in DWPF, and the less-radioactive constituents, still in the salt solution, would be stabilized with grout to create a saltstone waste form for onsite disposal as low-level radioactive waste.

In 1994 DOE selected the In-Tank Precipitation (ITP) process to separate the high-activity fraction from the salt solution. This process was designed to be carried out primarily in one of the underground HLW storage tanks with a 1.3 million-gallon (4,920 m^3) capacity. An inorganic sorbent (MST) was selected to remove radioactive actinides and strontium from the salt solution. An organic reagent (sodium tetraphenylborate) was selected to precipitate radioactive cesium from the salt solution. The ITP process included washing and filtration steps to separate the solid phases that contain these radioactive materials.

Sodium tetraphenylborate, the reagent used to precipitate cesium in the ITP process, is subject to catalytic and radiolytic decomposition. Decomposition of tetraphenylborate returns the cesium to the salt solution and results in the generation of benzene. Benzene is a toxic, flammable, and potentially-explosive organic substance that must be safely controlled. To achieve the objectives of the ITP process, the decomposition of tetraphenylborate has to be controlled as follows:

1. The amount of precipitated cesium that is returned to the salt solution has to be minimized.
2. The amount of generated benzene has to be significantly reduced.

The ITP process was designed to accommodate some tetraphenylborate decomposition, and to limit benzene accumulation. In the 1980s, SRS developed a salt treatment process, called ITP, and demonstrated it in moderate and full scale with actual SRS waste. During radioactive startup in 1995, ITP produced benzene in quantities significantly higher than predicted. The Westinghouse Savannah River Company (WSRC), the operating contractor, initiated laboratory and facility tests to determine the cause of escalated benzene generation and to manage it safely. In August 1996, the Defense Nuclear Facilities Safety Board (DNFSB, an independent oversight board for DOE) recommended that the ITP facility suspend operations until the process of benzene generation, retention, and release were better understood. The WSRC evaluated these processes from August 1996 through March 1998, and in January 1998, advised the U.S. Department of Energy-Savannah River Operations Office (DOE-SR) that the existing system configuration could not cost-effectively meet safety and production requirements for the ITP facility. In February 1998, the DOE Assistant Secretary for Environmental Management (EM-1) approved a DOE-SR plan to suspend startup-related activities and conduct a systems engineering study of alternatives to ITP.

TECHNOLOGY SELECTION

Technology Down-Selection Process

As described above, the U.S. Department of Energy (DOE) determined that In-Tank Precipitation (ITP) startup should be suspended and that a new cesium removal technology should be selected. The Westinghouse Savannah River System (WSRC) chartered Systems Engineering Team to evaluate potential cesium removal technologies in terms of process engineering; chemistry; waste processing; science; safety and regulatory engineering; operations; and systems engineering and interface controls. Members of the Systems Engineering Team also provided perspectives from industry, international radioactive waste disposal programs, and other DOE facilities that have large radioactive waste disposal programs. The Savannah River Technology Center (SRTC) and other DOE national laboratories—including Oak Ridge National Laboratory (ORNL) and Argonne National Laboratory (ANL)—provided research and development support. Several universities also provided needed expertise in specific areas. The Systems Engineering Team performed extensive evaluation of the problem and performance requirements and, from this, identified over 140 technologies that had some potential to address the problem.

The systems engineering study (WSRC 1998) evaluated over 140 technologies and reduced the list of candidates to four alternative cesium removal processes: Crystalline Silicotitanate Non-Elutable Ion Exchange (CST); Caustic-Side Solvent Extraction (CSSX); Small Tank Tetraphenylborate Precipitation (STTP); and direct grouting (which did not include cesium removal). During this study, the WSRC System Engineering Team performed several key aspects of the technology decision process, including definition of requirements for the Salt Processing Project (SPP). Table 2 lists the Key Functional Requirements for the technology evaluated. These requirements constituted a viability screening-process, and any technology that could not meet these requirements was disqualified from further consideration.

As research and development on the three cesium removal technology alternatives progressed, DOE conducted independent evaluations to make sure that their technical maturity was adequate to support selection.

These recommendations prompted the DOE Under Secretary and Assistant Secretary for Environmental Management to direct additional research and development on each alternative to reduce the technical uncertainty prior to down-selection.

Table 2. Key Functional Criteria (Harmon et al. 2001a).

Focus Area	Functions
Safety	
Hazard Assessment (HAD)	Provide a facility that meets the requirements of a non-reactor nuclear hazard category 2 and low chemical hazard category.
Interface Streams	
DWPF Recycle	Support tank farm space management and DWPF recycle evaporator strategy.
DWPF Glass	Provide a Cs-containing product that supports glass waste form requirements relative to durability, crystallization temperature, Na content, and viscosity.
SPF Feed	Provide a Decontaminated Salt Solution (DSS) product that meets Waste Acceptance Criteria relative to producing a non-hazardous saltstone waste form suitable for disposal as low-level solid waste at the SRS.
Tank 49H	Support Tank Farm space management strategy to recover Tank 49H for HLW storage.
Tank 50H	Support Tank Farm space management strategy to recover Tank 50H for HLW storage.
New Waste Form	Comply with DOE-RW HLW repository requirements.
Nominal DF	
Sr DF	Provide a Sr DSS concentration of $\leq$ 40 nCi/g, which equals a nominal DF = 5 (overall average).
TRU DF	Provide a TRU DSS concentration of $\leq$ 18 nCi/g, which equals a nominal DF = 12 (overall average).
Cs DF	Provide a Cs DSS concentration that enables conversion to a solid low-level waste form suitable for near-surface disposal at the SRS. Cs-137 $\leq$ 45 nCi/g is required to enable processing in the existing Saltstone Production Facility and disposal in the existing Saltstone Disposal Facility, which equals a nominal DF = 8000 (overall average).
Schedule	
HLW Storage	Support Tank Farm space management strategy to support site missions (timely startup of new process by 2010).
FFA	Support readiness for closure of all waste tanks by 2028.
STP	Support readiness for closure of old style tanks by 2020, and an average glass canister production rate of 200 canisters/yr.

The DOE established a Technical Working Group (TWG) to manage the technology down-selection process. In March 2000, DOE requested the Tanks Focus Area (TFA) to manage the SPP technology development program for the TWG and to support a technology selection by June 2001. The TFA was requested to review and revise the technology roadmaps, and prepare a comprehensive research and development program plan for the three candidate cesium-removal technologies as well as the actinide and strontium-removal processes that are part of the overall SPP. In addition, TFA was requested to develop a decision process and down-selection criteria as described below.

Decision Analysis Approach

The SPP decision process (Harmon et al. 2001b) focused its analysis on three levels: 1) identification of goals that the selected technology should achieve, 2) selection criteria that are a measure of performance of the goal, and 3) criteria scoring and weighting for each technology alternative. Six goals were identified that appeared to be clear, unique, and inclusive objectives for the project: 1) meet schedule; 2) minimize cost; 3) minimize technical risk; 4) minimize environmental safety and health risk; 5) minimize impact to interfaces; and 6) process flexibility. The technology alternatives were analyzed and criteria were defined that reflected how well each technology achieved the project goals.

The DOE goals for the process selection at SRS were:

1. Meet schedule (Criterion 1)
2. Minimize cost (Criteria 2 and 3)
3. Minimize technical risk (Criteria 4 and 5)
4. Minimize environmental safety and health impacts (Criteria 6 and 9)
5. Minimize impact to interfaces (Criteria 7 and 8)
6. Maximize process flexibility (Criteria 10 and 11)

A series of trial scoring, sensitivity analysis, and weighting were conducted using these criteria to confirm their efficacy. The effectiveness of the initial criteria was based on the following factors:

1. Differentiation between technology alternatives
2. Relationship to goals or values of DOE and other stakeholders
3. Measurable or estimable
4. Reasonably independent of each other
5. Well understood by all decision makers

Table 3 summarizes the final criteria and definitions that were approved by DOE Headquarters in August 2000. In the scoring process, the specific scores assigned varied from 1 to 5 for each criterion, with 1 being the worst, and 5 being the best. The Project Team also assigned weights to the goals, which were then allocated among the criteria associated with each goal, to provide a measure of their relative significance and confidence. The scores then multiplied the weights for each criterion, with these products being summed for each technology to obtain a final score for each technology (Fig. 3). The Project Team also considered other factors, including strengths and weaknesses of each alternative and implementation characteristics of each technology such as reliability, maintainability, inspectability, and recoverability from process upsets.

Independent Reviews and DOE's Response

In February 2000, DOE decided to evaluate the four candidate alternatives for replacement of the ITP process, and to conduct an extensive research and development program on the three alternative cesium removal technologies (Direct Disposal in Grout did not require cesium removal). The report of the National Research Council (NRC 2000)—the research arm of the National Academy of Sciences, National Academy of Engineering, and Institute of Medicine—endorsed in general the selection of the four candidate processes considered as alternatives for salt disposal, concluding that each of the processes was potentially appropriate and no obvious major processing options were overlooked. The NRC report identified technical uncertainties associated with each of the alternatives, as well as schedule constraints and potential regulatory restrictions.

Table 3. Down Selection Criteria and Definitions (Harmon et al. 2001b).

Criterion No.	Definition and Description
1.	**Schedule technical risk:** This criterion addresses risk to the overall project schedule due to high-risk technology issues not being resolved in time to support down-selection.
2.	**Project cost reduction potential:** Potential that cost-savings in the total project cost can be identified, generally due to flowsheet or equipment arrangement changes that would allow facility footprint reductions.
3.	**Life-cycle costs through decontamination and decommissioning:** Total costs to complete all salt processing (including HLW system costs). The focus is on life-cycle costs, but the separate components' total project cost and operating cost are also examined for key differences.
4.	**Technical maturity:** This criterion addresses the overall technical maturity of the process flowsheets, including the required strontium and actinide removal steps. Stages of maturity as provided by the gate system of the Office of Science and Technology are applied to each unit operation and the results are averaged.
5.	**Implementation confidence:** This criterion deals with the amount of relevant process experience, particularly large-scale demonstration or deployment in the DOE complex and industry for the key equipment used for each cesium removal process. This criterion also includes commercial availability of key components and chemicals.
6.	**Minimize environmental impacts:** A comparative assessment of environmental impacts from secondary waste streams, airborne emissions, and liquid effluents is included here. This criterion also includes the number of Saltstone vaults required for each process.
7.	**Impacts of the interfaces at the DWPF:** This criterion evaluates cost of implementing the changes (physical modifications) to the interfacing systems and the loss of glass canister production caused by outages for equipment installation or transfer line tie-ins.
8.	**Process simplicity to interfacing systems:** The simplicity of interfacing the alternative cesium removal processes with other high-level waste systems is assessed in this criterion. Note that the simplicity is measured by the number of process unit operations needed for the interface times—a difficulty factor for each interface unit operation.
9.	**Levels of safety control mitigation:** This criterion addresses the number and type (e.g., passive, active, administrative, preventive, and mitigative) of controls required to maintain the facility in a safe configuration and to protect the worker, public, and environment.
10.	**Maximize process flexibility in throughput:** The capability to operate the process at a higher or lower throughput (turn-up or turn-down) based on the equipment in the current pre-conceptual designs is addressed in this criterion.
11.	**Maximize process simplicity (operability):** This criterion addresses the simplicity of the process as indicated by the number of pieces of equipment (in both the non-radioactive areas and the remotely-operated area) and number of jumpers (piping connections) required inside the remotely-operated area.

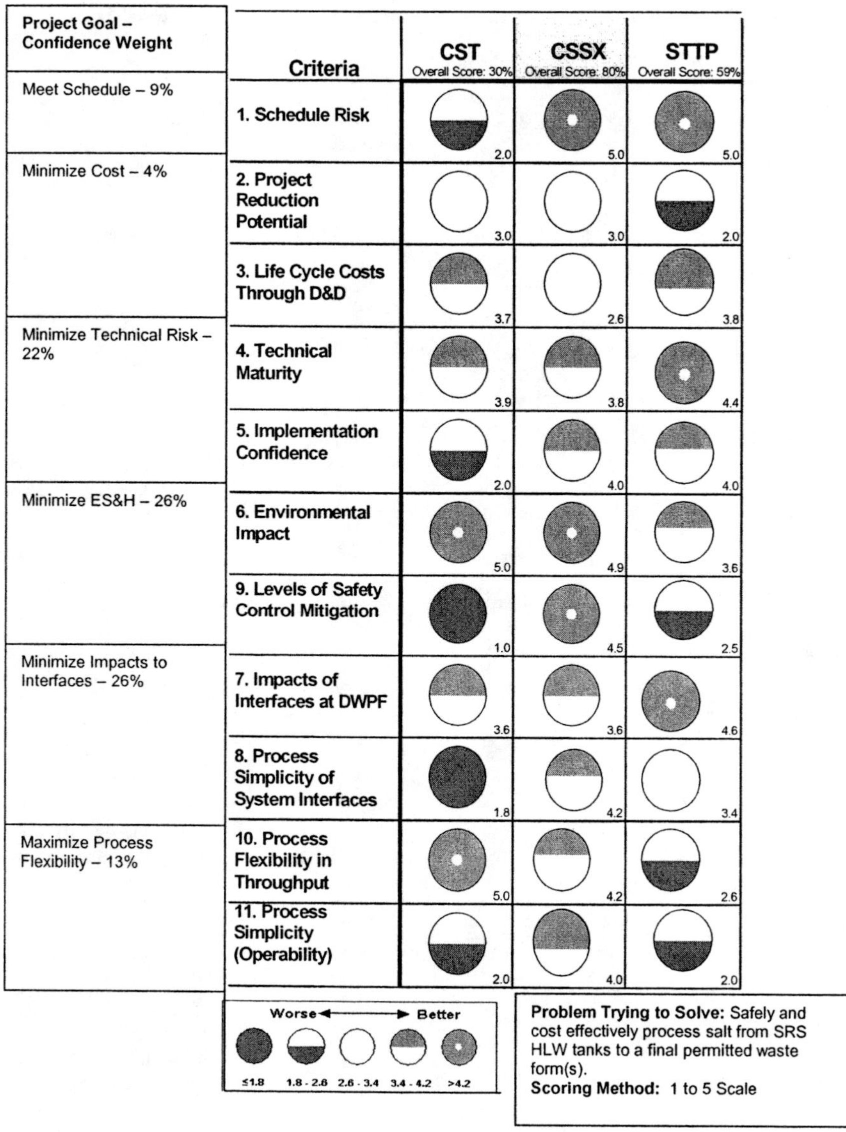

Fig. 3. TWG's Final Evaluation (Harmon et al. 2001b).

In connection with the first NRC report, DOE asked the NRC to provide a follow-on assessment of DOE's efforts to evaluate and select a process for separating radionuclides from soluble high-level radioactive waste at the Savannah River Site (SRS). A second NRC committee was appointed and tasked to:

1. evaluate the adequacy of the criteria DOE used to select from among the candidate processes under consideration.
2. evaluate the progress and results of the research and development work DOE undertook on the candidate processes.
3. assess whether the technical uncertainties were sufficiently resolved to proceed with downsizing the list of candidate processes.

The resulting interim report (NRC 2001a) concluded that: 1) DOE's selection criteria were reasonable and appropriate and were developed in a transparent way; 2) Some of the criteria did not appear to be independent of others; and 3) Some criteria appeared unlikely to discriminate among the process alternatives. The NRC recommended that these criteria should not be implemented in a way that relies on a single numerical total score, but instead: "Rather than averaging and totaling the scores for each criterion, the various criteria should be seen as relevant to different goals and purposes and should be considered individually. Some of the criteria should be used as 'go/no go' gates and some should have thresholds for use."

In its final report, the NRC (2001b) concluded "that technical uncertainties have been resolved sufficiently to proceed with downselecting the list of candidate processes." The report also noted that "[u]nless tests with actual waste encounter new problems, the CSSX option for cesium separation presents, at present, the fewest technical uncertainties of any of the three cesium separation alternatives."

After completion of the laboratory research and development activities, DOE had to evaluate the results of the studies in order to identify the best technology among different competing technologies. In order to produce the final design of the facility implementing the full-scale operation of the selected technology, DOE may have to construct and operate a Pilot Plant of this selected technology (DOE 2001).

Timing of Technology Selection

Although any alternative process to ITP would be specifically developed to enable HLW salt disposition, the impact on all HLW facilities and processes at SRS must also be addressed. Functionally, the selected alternative must interface safely and efficiently with the processing facilities within and outside the HLW System. In determining the timing for the selection of an alternative, consideration should be given to tank farm space and water inventory management; the Site Treatment Plan (STP); and the Federal Facility Agreement (FFA) for tank closure.

There are reasons to believe that if salt processing is delayed beyond 2010, the salt waste must be vitrified separately from the sludge component of the HLW, and the total number of HLW canisters would be increased over the number projected for concurrent sludge and salt waste vitrification. Preliminary projections indicate that if the salt processing initiation date of 2010 is not met, then the potential exists that up to 150 additional canisters (salt-only) per year would have to be produced for every year in which startup is delayed beyond 2010. The cost for additional canister production would be about $300 million per year. In the event that sludge processing is completed prior to the initiation of salt processing, it would take 13 years (at 150 canisters per year) to process all the salt waste at an approximate cost of $4 billion in addition to the cost of construction and operation of the salt-processing facility. These costs do not include federal repository costs for transportation and disposal (DOE 2001).

Environmental Impacts of Salt Processing Alternatives

The selection of technologies for processing HLW salt waste at SRS required an environmental assessment of various alternatives. This assessment included the comparative evaluation of potential adverse human health and ecological impacts for each alternative. Additionally, this assessment had to consider the technical feasibility and regulatory acceptability of each alternative. Recognizing the complexity of the subject, DOE sought the advice of several organizations on various aspects of the selection process. Subsequently, DOE made a final decision considering various alternatives including the alternative of no action.

Common features of all processing alternatives include initial separation of low-concentration soluble radioactive strontium and actinides (including plutonium) by sorption (Table 4) onto granular solid MST, followed by filtration. The essential difference among the alternatives is the technology for removal of the relatively high concentrations of radioactive cesium, except for the Direct Disposal in Grout alternative, in which cesium would not be removed. The final waste forms are similar for each of the other action alternatives with the high-activity salt fraction extracted from the salt and incorporated into the vitrified waste form for eventual repository disposal; and the low-activity salt fraction immobilized as saltstone for onsite disposal. In the Direct Disposal in Grout alternative—although the high-activity salt solution would also be vitrified for eventual repository disposal—the cesium would remain in the fraction immobilized as saltstone for onsite disposal.

Table 4. Comparison of Salt Processing Alternatives (DOE 2001).

	Process stages			
Salt processing alternatives	Strontium and actinide (Pu) removal from salt solution	Cesium removal from salt solution	Final waste form	
			DWPF glass (HLW)	Saltstone (LLW)
Small Tank Precipitation	MST sorption	TPB Precipitation	MST/TPB solids	Low-activity salt solution
Ion Exchange	MST sorption	CST Ion Exchange	MST solids, CST resins	Low-activity salt solution
Solvent Extraction	MST sorption	Organic extractant	MST solids, aqueous cesium solution	Low-activity salt solution
Direct Disposal in Grout	MST sorption	None	MST solids only	Cesium salt solution

MST = Monosodium Titanate, CST = Crystalline Silicotitanate, TPB = Tetraphenylborate, HLW = high-level waste, LLW = low-level waste.

A diagram comparing the process phases for the salt processing alternatives are presented in Fig. 4. Potential Salt Waste Processing Facility sites are shown in Fig. 5.

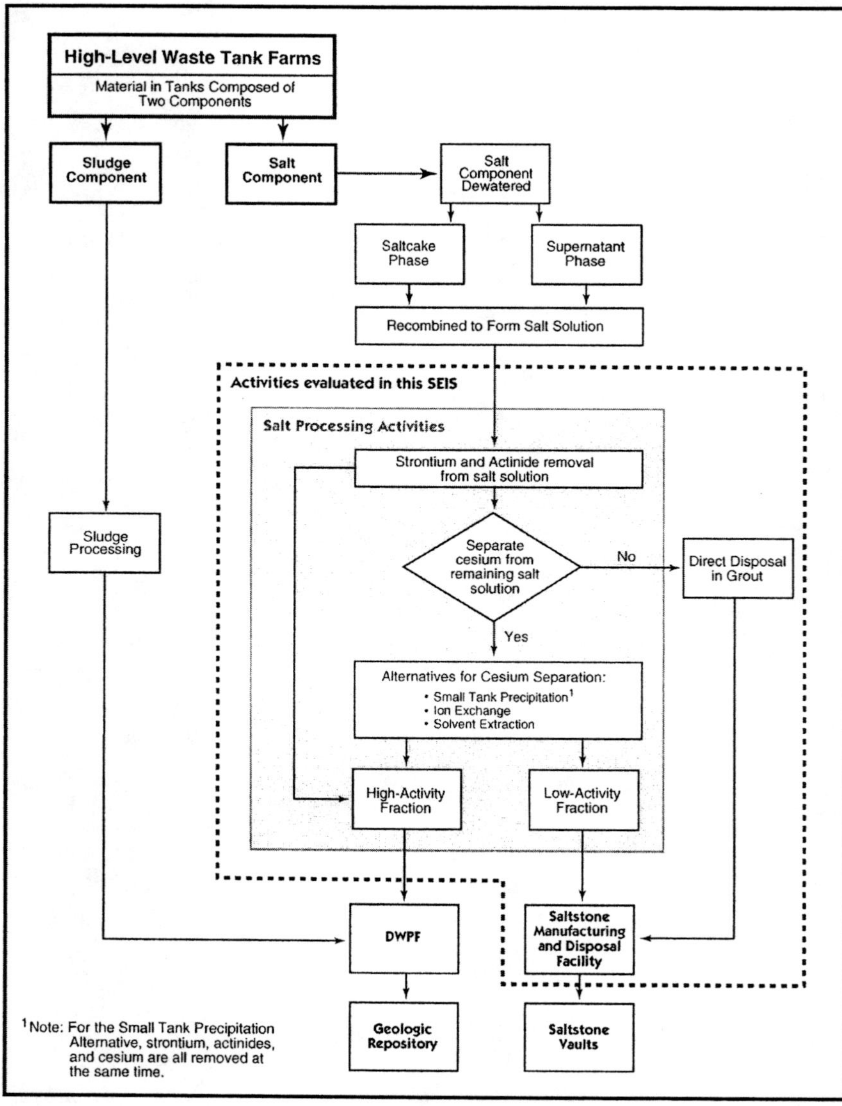

Fig. 4. Process Flow for High-Level Waste at the Savannah River Site (DOE 2001).

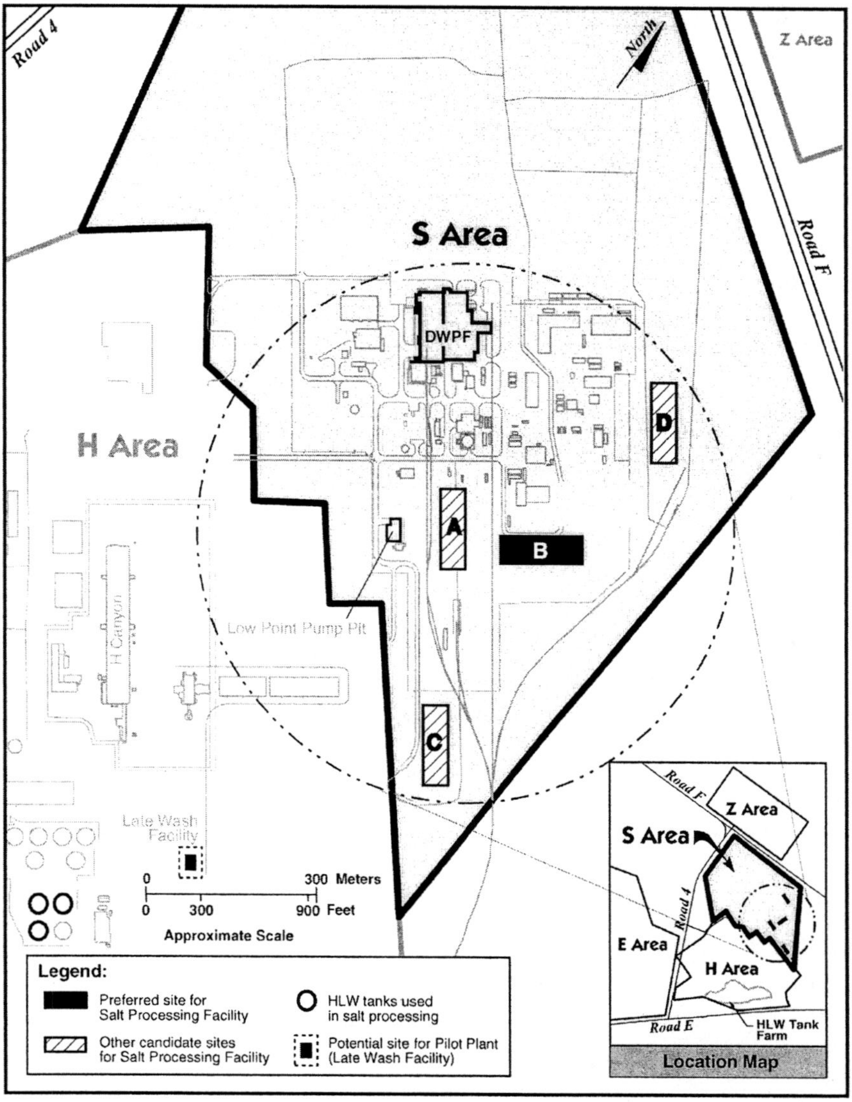

Fig. 5. Potential Salt Processing Facility Sites (DOE 2001).

Small Tank Precipitation: The Small Tank Precipitation alternative would use tetraphenylborate precipitation, the same chemical reaction as ITP, to remove the radioactive cesium from the HLW salt solution.

However, the process would be conducted as a continuous operation using a small, temperature-controlled reaction vessel to reduce or inhibit the tetraphenylborate decomposition and benzene generation. The vessel and operating conditions would be designed to minimize benzene emissions and flammability hazards by maintaining an inert gas (nitrogen) atmosphere within the reaction vessel. Radioactive cesium would be separated from the salt solution by precipitation as an insoluble tetraphenylborate solid. Radioactive actinides and strontium would be removed concurrently by sorption onto a granular solid (MST). These solids would be separated from solution and concentrated by filtration, then treated chemically to decompose the tetraphenylborate precipitate and remove the benzene formed. The solids slurry containing the separated radioactive constituents would be transferred to the Defense Waste Processing Facility (DWPF) for vitrification. The low-activity salt fraction would be transferred to the Saltstone Production Facility (SPF) for disposal as grout in onsite vaults.

Ion Exchange: The Ion Exchange alternative would use crystalline silicotitanate resin in ion exchange columns to separate cesium from the salt solution. The salt solution would pass through large stainless steel ion exchange columns filled with the ion exchange resin to react the cesium with the resin. Treatment of the solution with monosodium titanate to separate strontium and actinides, and filtration to remove those solids and residual sludge, would be necessary prior to separating the cesium to prevent plugging the ion exchange columns. Both the MST solids and the cesium-loaded crystalline silicotitanate resin would be transferred to DWPF for vitrification. The low-activity salt solution would be transferred to the SPF for disposal as grout in onsite vaults. The Ion Exchange process would result in the accumulation of as much as 15 MCi of ^{137}Cs on the resin inventory within the process cell. This radioactive loading would necessitate stringent shielding requirements and operational controls because of high radioactivity, high heat generation, and the generation of hydrogen and other gases.

Solvent Extraction: The Solvent Extraction alternative would use a highly-specific organic extractant to separate cesium from the HLW salt solution. The cesium would be transferred from the aqueous salt solution into an insoluble organic phase, using a centrifugal contactor to provide high surface area contact, followed by centrifugal separation of the two phases. Recovery of the cesium by back extraction from the organic phase into a secondary aqueous phase would generate a concentrated cesium solution for vitrification in DWPF. Before separating cesium from the salt solution, MST would separate soluble actinides and strontium followed by filtration to remove the solids and residual sludge. This pretreatment would be required to meet salt solution decontamination requirements and avoid interference in the solvent extraction process. The MST solids would be transferred to DWPF for vitrification along with the concentrated cesium solution. The low-activity salt solution would be transferred to the SPF for disposal as grout in onsite vaults.

Direct Disposal in Grout: Along with the high-activity fraction, the first three technologies presented above remove cesium from the salt solution for eventual disposal as HLW. Under the Direct Disposal in Grout alternative, the HLW salt solution would be disposed onsite as saltstone without separation of radioactive cesium. Prior to solidifying the salt solution as grout, MST would be used to remove the strontium and actinides to meet saltstone waste acceptance criteria as LLW. The MST slurry would be transferred to DWPF for incorporation into HLW glass. After the MST treatment, the clarified salt solution would be combined with flyash, cement, and slag in a grout mixer for disposal in the saltstone vaults. The resulting saltstone would have radionuclide concentrations less than Class C LLW, but would exceed Class A limits, as defined by the USNRC (2003). These waste classifications are not generally applicable to DOE-generated LLW. However, the current saltstone permit—which was issued by the South Carolina Department of Health and Environmental Control (SCDHEC) under its State wastewater authority—authorizes disposal of wastes with radionuclide concentrations comparable to Class A LLW. Under the permit, DOE must notify SCDHEC if the characteristics of wastes in saltstone vaults would change, as would be the case with the higher level of radioactivity in the final waste form under the Direct Disposal in Grout alternative.

No Action Alternative: Under the No Action alternative in the near-term, DOE would continue current HLW management activities—including tank space management—without a process for separating the high-activity and low-activity salt fractions. The DWPF would vitrify only sludge from the HLW tanks. Saltcake and salt supernatant would be stored in the HLW tanks and monitoring activities would continue. Tank space would continue to be managed to ensure adequate space to meet safety requirements and closure commitments. Current tank space management projections indicate that after 2010, additional tank space would be needed to support continued operations under the No Action alternative. The DWPF operations result in large volumes of waste, mostly water, that is returned to the HLW tanks. DOE uses evaporators to substantially reduce this volume but, until a salt processing alternative is on-line, DWPF operation will increase rather than decrease the volume of HLW that must be stored in the tanks. To maintain tank space until about 2010, tank space management under the No Action alternative would require numerous initiatives, potentially including suspension of DWPF operations in order to provide storage capacity in the HLW tanks.

A summary of the long-term impacts of the salt processing alternatives is shown in Table 5.

DOE's Decision

DOE has performed research on each of the three cesium-removal technology alternatives to establish the technological risk involved in implementing each alternative. The final conclusion supported the technology selection and the identification of the preferred alternative. The Final Supplemental Environmental Impact Statement (DOE 2001) assessed the potential environmental impacts of each alternative. Additionally, DOE conducted a final management review that comparatively evaluated each of the alternatives against a list of criteria that included cost, schedule, technical maturity, technology implementability, environmental impacts, facility interfaces, process simplicity, process flexibility, and safety. On the basis of this final review and subsequent DOE senior management validation, DOE selected MST strike with filtration for removal of actinides, and strontium and CSSX for cesium removal as the technologies to be implemented in the planned Salt Waste Processing Facility (SWPF).

Table 5. Summary Comparison of Long-Term Impacts by Salt Processing Alternative. Bold indicates the alternative with the greatest impact for a particular parameter (DOE 2001).

Parameter	No Action	Small Tank Precipitation	Ion Exchange	Solvent Extraction	Direct Disposal in Grout
Nitrate concentration at 100-m well (mg/L)[a]	NA	29	31	26	**33**
Radiation dose (mrem/y) from 100-m well	**640**[b]	0.042	0.044	0.038	0.048
LCF from 100-m well[c]	**0.022**[b]	1.5×10^{-6}	1.5×10^{-6}	1.3×10^{-6}	1.7×10^{-6}
Radiation dose from Agricultural Scenario (mrem/y)	NA	110	130	110	**140**
LCF from Agricultural Scenario[c]	NA	3.9×10^{-3}	4.6×10^{-3}	3.9×10^{-3}	$\mathbf{4.9\times10^{-3}}$
Radiation dose from Residential Scenario at 100 y post-closure (mrem/y)	**2,320,000**[d]	0.11	0.13	0.1	1,200[e]
LCF from Residential Scenario at 100 y post-closure[c]	1.16[f]	3.9×10^{-6}	4.6×10^{-6}	3.5×10^{-6}	$\mathbf{4.2\times10^{-2}}$
Radiation dose from Residential Scenario at 1,000 y post-closure (mrem/y)[g]	NA	69	80	65	**85**
LCF from Residential Scenario at 1,000 y post-closure	NA	2.4×10^{-3}	2.8×10^{-3}	2.3×10^{-3}	$\mathbf{3.0\times10^{-3}}$

a. Nitrate MCL is 10 mg/L.

b. Based on consumption of contaminated surface water at Fourmile Branch.

c. Health effects are expressed as lifetime (70 y) individual probability of a latent cancer fatality.

d. Based on external radiation in the area of the tank farm.

e. The external dose for direct disposal in grout alternative in the 100-y scenario is primarily due to ^{137}Cs (half-life 30 y). For all other alternatives and scenarios, the external dose is primarily due to the isotopes with long half-lives.

f. Probability of an LCF provided for comparison. The external radiation dose from No Action would result in prompt fatalities.

g. External radiation doses at 1,000 y post-closure are higher than doses 100 y post-closure because a layer of soil to provide adequate shielding is assumed to be present in the 100 y scenario, but is assumed to be absent in the 1,000 y scenario.

LCF = latent cancer fatality.

NA = not applicable.

Solvent Extraction was selected as the salt processing alternative primarily because it presents the least technical risk for successfully removing cesium from radioactive waste. Although Solvent Extraction uses a complex four-component solvent system, laboratory testing has shown that component concentration and process flow can be maintained to effectively remove cesium from the wastes. Other key strengths identified for the Solvent Extraction technology include the following:

1. Maturity of and experience within the DOE Complex for processing nuclear material.
2. Simplicity with which the Solvent Extraction product stream could be incorporated into the current Defense Waste Processing Facility vitrification process.
3. The ability to rapidly start-up and shut-down the Solvent Extraction centrifugal contactors.

Solvent Extraction is comparable to the other action alternatives with regard to its short-term and long-term environmental impacts.

Following the technology selection in 2001, extensive technology development activities for CSSX and MST/Filtration technologies continued at several universities and DOE laboratories in an effort to mature these technologies and to optimize them for use in the SWPF. In 2002, DOE selected two Engineering, Procurement, and Construction (EPC) contractors to develop competitive conceptual designs for the SWPF based on CSSX and MST/Filtration technologies. In January 2004, DOE plans to complete SWPF's conceptual design and request project Critical Decision One (CD-1, Approval of Alternative Selection and Cost Range). Beyond CD-1, a single EPC contractor will complete design, construction, and commissioning of the SWPF.

MONOSODIUM TITANATE

The current conceptual design for the SWPF requires removal of strontium and transuranic radionuclides in advance of removing cesium from the solution. The selected technology involves addition of an inorganic sorbent, MST, and subsequent removal of solids by cross-flow filtration. The MST shows a high affinity for strontium and also effectively removes soluble actinides such as plutonium and uranium from solution. The MST also sorbs lesser amounts of neptunium and other alpha-emitting radionuclides. The treated liquid (filtrate) is processed by solvent extraction to remove cesium. The collected solids require washing to reduce the concentration of soluble salts of sodium prior to transfer to the DWPF. Primary radionuclide removal requirements for SWPF are given in Table 6.

Process Description

A salt solution feed batch at 6.44 M Na^+ will be transferred to the Alpha Sorption Tank (AST) for dilution to ~5.6 M Na^+ (Fig. 6). Salt solution is received into the AST, where it is diluted and batch-treated with a slurry of MST to sorb plutonium, neptunium, uranium, and strontium onto the MST. After 24 hours of sorption time, the dilute MST/sludge slurry in the AST is transferred to the Filter Feed Tank (FFT). MST/sludge slurry will be fed continuously from the FFT to a cross-flow filter unit to remove both MST and sludge solids. The filtrate stream (clarified salt solution) will be sent to the CSSX Salt Solution Feed Tank.

Table 6. Radionuclide Removal Requirements (Dimenna et al. 2001).

Component	Saltstone WAC (nCi/g)	Required DF* Average/Bounding
Cesium-137	45	7,700/40,000
Plutonium/Americium	18 (total alpha)	12/49
Uranium	18 (total alpha)	1/1
Neptunium	0.03	NA/2
Strontium	40	1/5.1

*values change with feed blend strategy

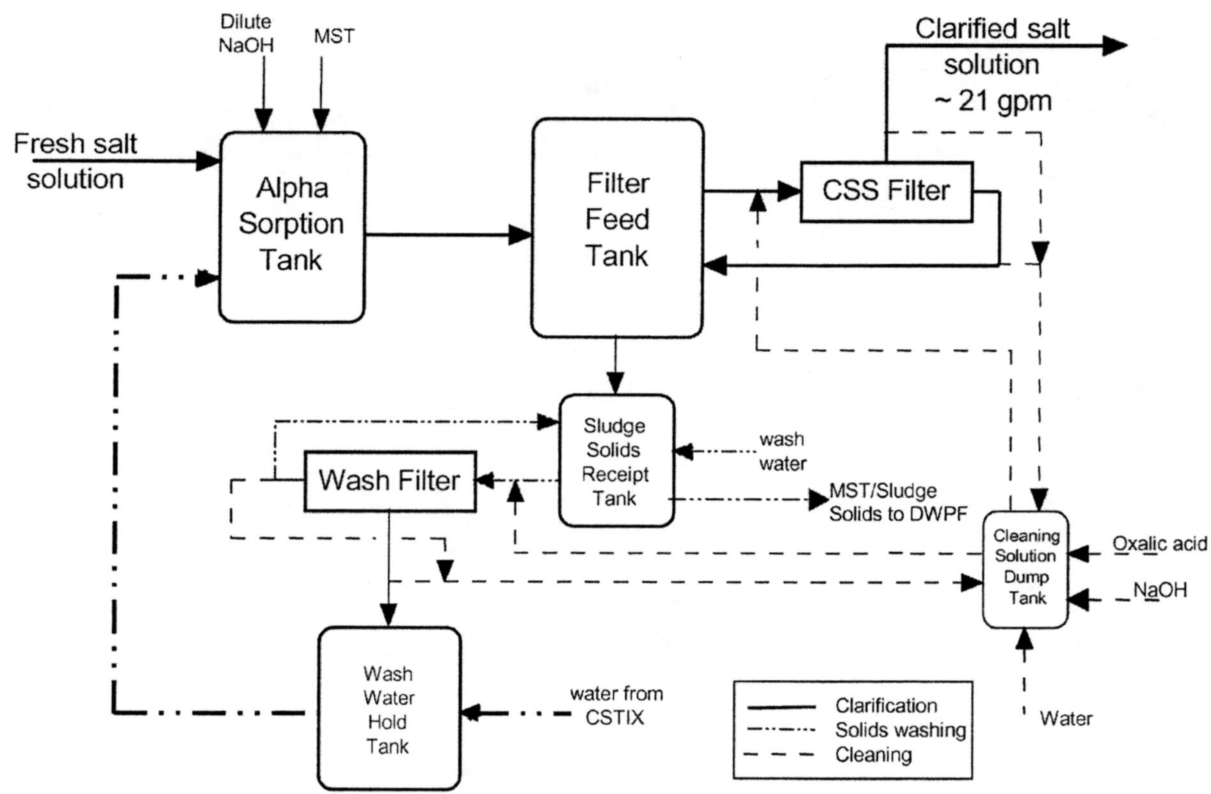

Fig. 6. Alpha Decontamination by MST Addition (Dimenna et al. 2001).

When the heel in the FFT is concentrated to about 5 wt % solids (five to eight salt solution feed batches), it will then be transferred to the Sludge Solids Receipt Tank (SSRT) for washing to reduce the sodium concentration in the solution associated with the MST/sludge solids to less than 0.5 M. A separate, smaller cross-flow filter will be used in the MST/sludge-washing cycle to avoid interrupting the production of clarified salt solution. The washed solids are then stored within the SWPF until they can be transferred to the DWPF for further processing. Cleaning solutions will be run through the cross-flow filter unit after two or three batches of salt solution have been processed through as CSSX feed to minimize fouling and plugging of the sintered metal filter media.

MST Concentration and Kinetics

The adsorption studies of strontium, plutonium, uranium, and neptunium on MST have been performed at MST concentrations of 0.2, 0.4, 1.1, and 2.0 g/L MST, and at sodium concentrations of 4.5 and 7.5 M. Strontium removal is fast for both sodium concentrations (Fig. 7), but both the extent and rate of plutonium removal is limited at 7.5 M (Fig. 8). Test results show that both the extent and rate of plutonium removal improve as the sodium concentration is reduced and as the MST concentration is increased (Fig. 9). The average salt solution is expected to be 6.44 M sodium. The sodium concentration chosen for the CSSX process is 5.6 M. To ensure removal of plutonium at average concentration and activity, the salt solution is diluted before the MST addition and the maximum MST concentration (0.4 g/L)—shown to be acceptable for DWPF glasses—is added.

A study of MST kinetics at 5.6 M sodium shows rapid strontium removal at a MST concentration of 0.4 g/L (Fig. 10). In addition, it shows that a 24-hour reaction period will reduce the soluble plutonium concentration to less than 20 μg/L (DF > 10) at this sodium concentration (Fig. 11). Since the soluble alpha activity in average waste is dominated by plutonium, these conditions ensure that the SPF alpha limit can be met after 24 hours of contact time for the overall average of blended salt solution fed to the CSSX process. However, the range of soluble plutonium concentrations in the Tank Farm (both total concentration and isotopic distribution) is not well characterized. Additional work may be required to ensure that sufficient plutonium can be removed at bounding conditions.

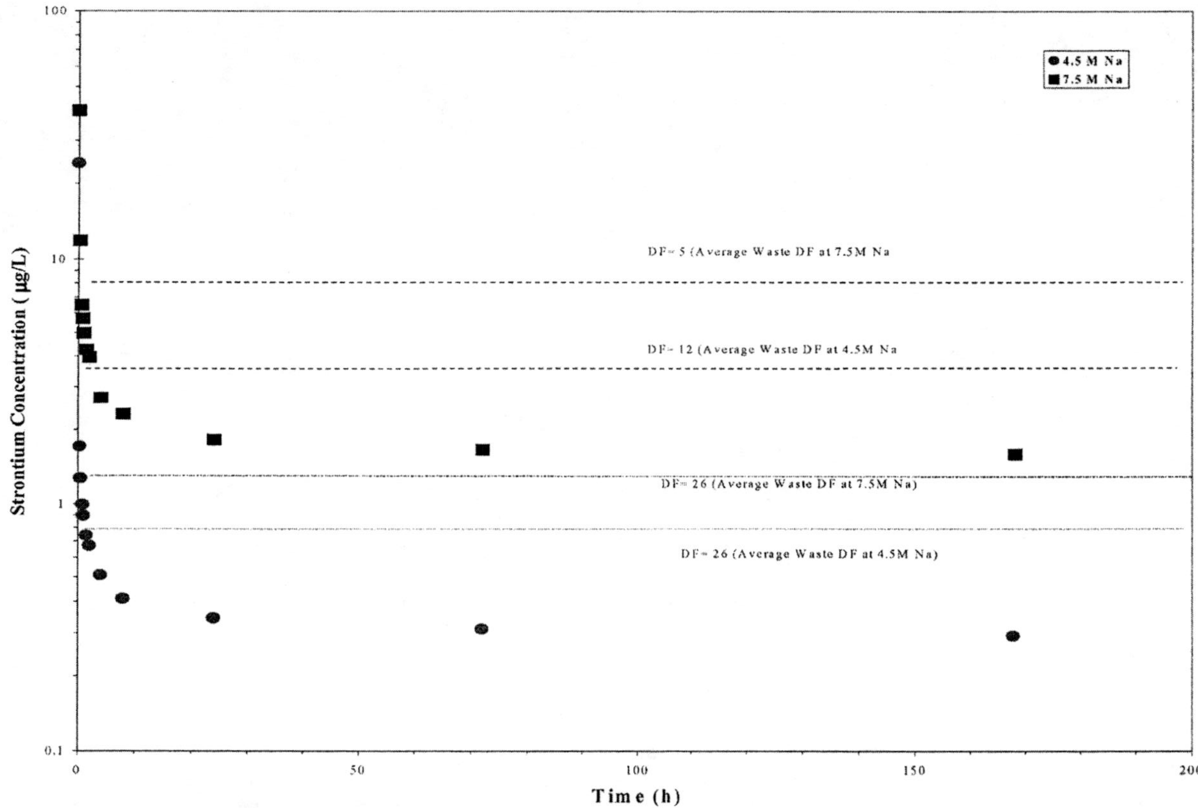

Fig. 7. Sr Removal Kinetics at 0.2 g/L MST Addition (Dimenna et al. 2001).

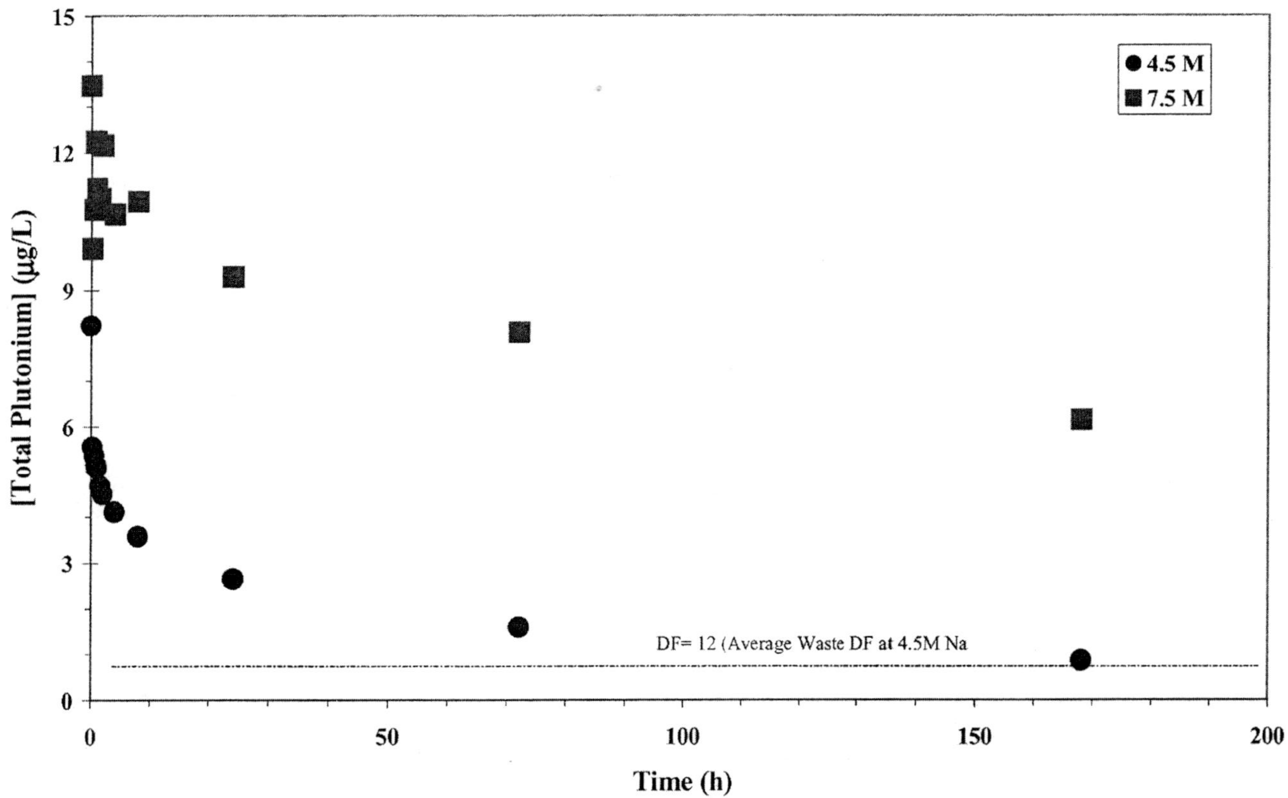

Fig. 8. Pu Removal Kinetics at 0.2g/L MST Addition (Dimenna et al. 2001).

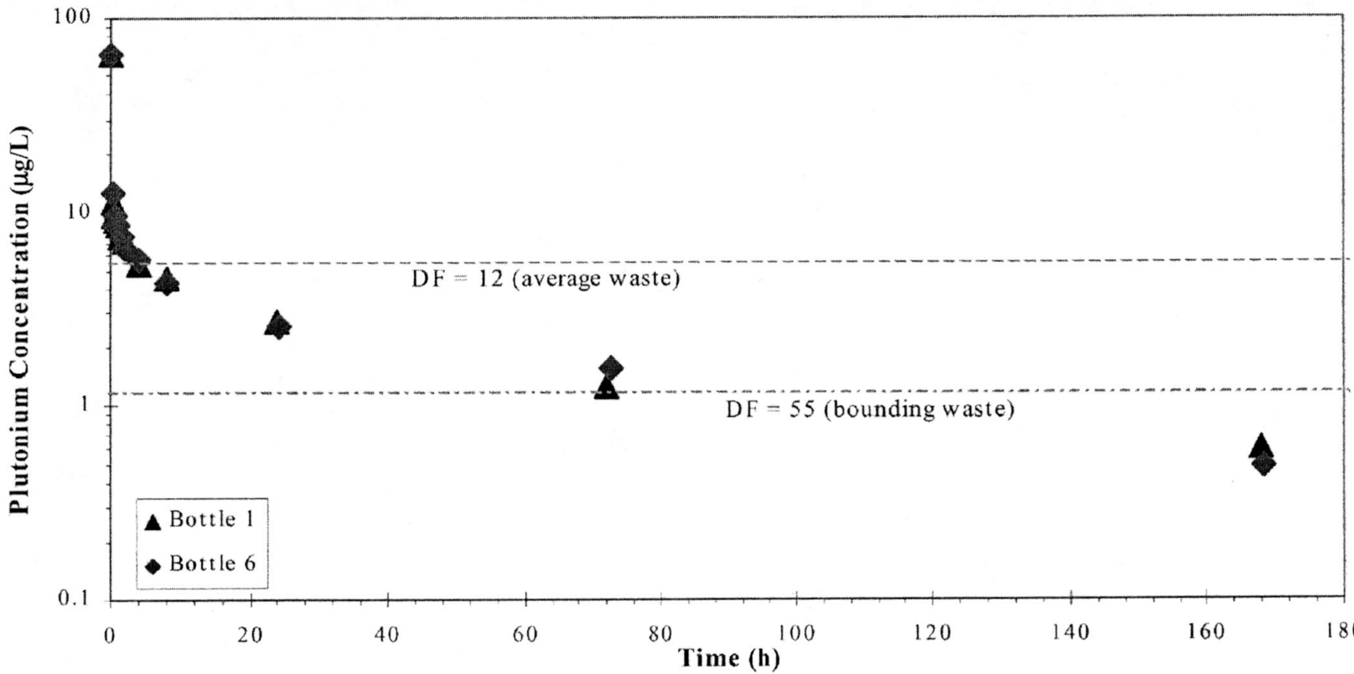

Fig. 9. Comparison of Plutonium Concentration Tests in 4.5 M Na and 0.4 g/L MST (Dimenna et al. 2001).

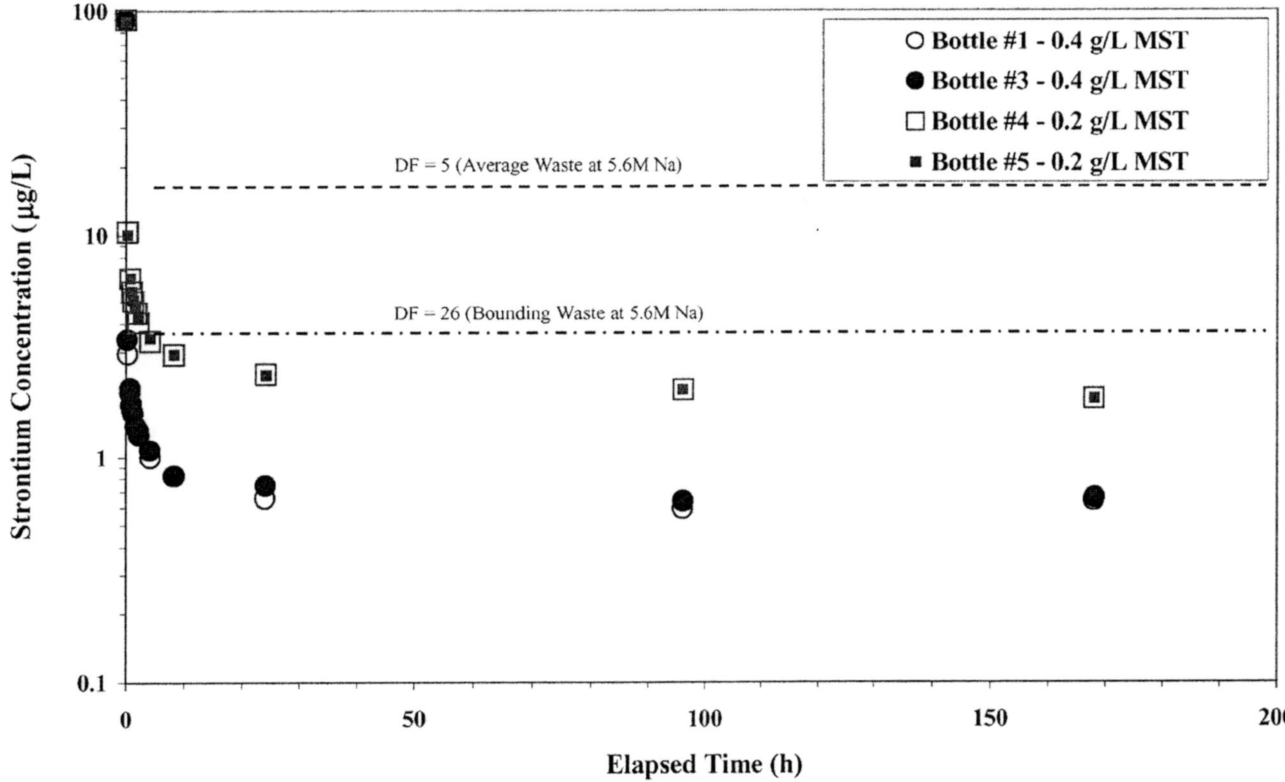

Fig. 10. Comparison of Sr Removal Kinetics 5.6 M Na at 0.2 and 0.4 g/L MST (Dimenna et al. 2001).

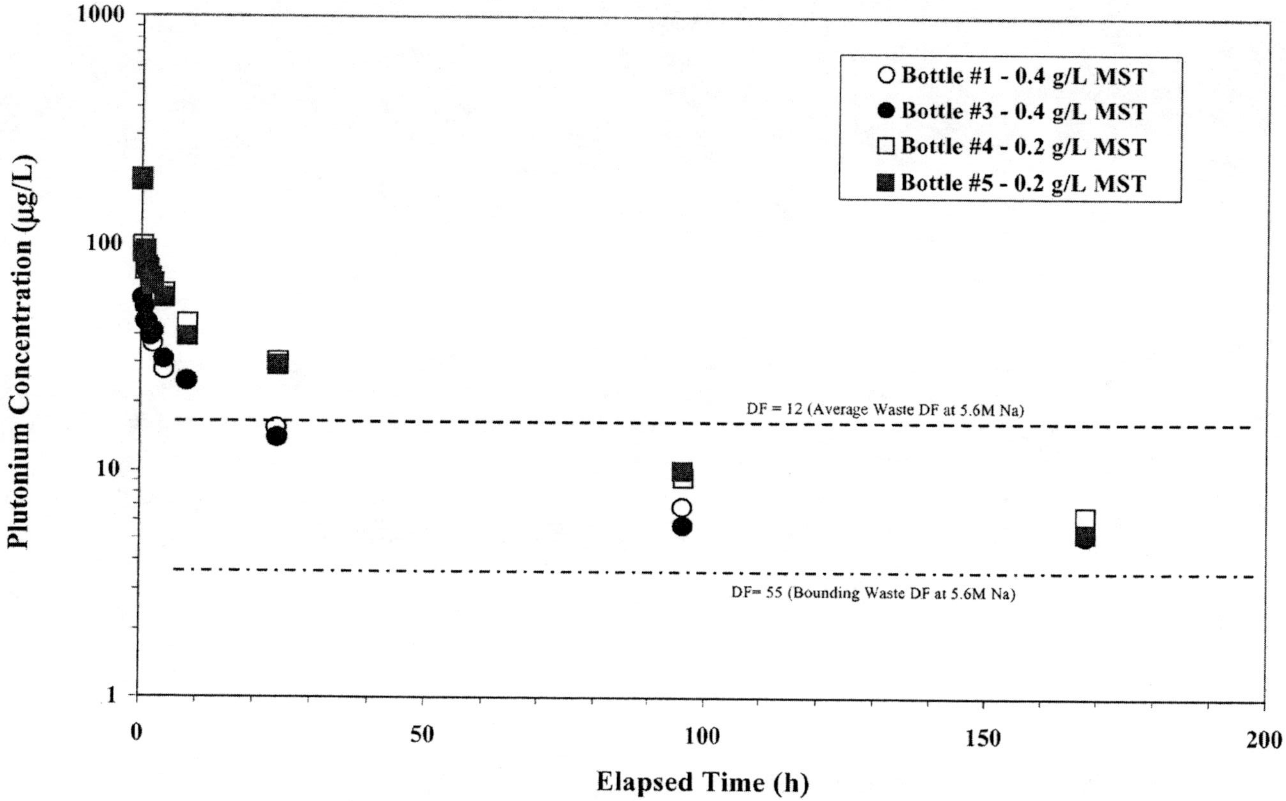

Fig. 11. Comparison of Pu Removal Kinetics in 5.6 M Na at 0.2 and 0.4 g/L MST (Dimenna et al. 2001).

MST Actual Waste Tests

The Project Team (Fink et al. 2002b, Harmon et al. 2002d) conducted several demonstrations of strontium and actinide removal with MST using tank waste. Primary objectives of this testing included: 1) confirm MST performance with actual tank waste solutions; 2) determine the effect of scale and mixing on MST performance; and 3) demonstrate sufficient removal of actinide and strontium to achieve Saltstone Waste Acceptance Criteria (WAC) with waste solutions at bounding radionuclide levels. For wastes high in ^{137}Cs, Savannah River Technology Center (SRTC) conducted the demonstrations in the SRTC Shielded Cells facility. For wastes low in ^{137}Cs, the testing was conducted in radiochemical hoods within the SRTC.

The SRTC received a large sample of waste from Tank 37H for demonstration of the actinide and strontium removal and the CSSX processes. Personnel diluted the Tank 37H material to provide 70 liters of the waste solution having a sodium ion concentration of 5.6 M. Initial ^{90}Sr and plutonium concentrations measured 1340 and 1.57 nCi/g, respectively. Actinide and strontium removal occurred in a large stainless steel reactor at ambient temperature with 0.4 g/L of MST. A central mechanical agitator provided mixing during the actinide and strontium removal process.

Decontamination factors for ^{90}Sr and plutonium measured 2.4 and 2.9, respectively. Both values are below that measured with simulated wastes. Note that the total plutonium activity of the diluted Tank 37H material (1.57 nCi/g) proved below the Saltstone WAC for total alpha activity (18 nCi/g) and, therefore, this waste solution did not require plutonium removal. The treated solution did not meet the Saltstone WAC for ^{90}Sr. The Project Team attribute the low strontium removal to two primary factors: 1) high ^{90}Sr content and 2) use of a low-capacity batch of MST.

Analysis of the diluted waste solution measured ^{90}Sr concentration of 1340 nCi/g, which is a factor of 7 times the estimated average value for ^{90}Sr. Thus, the Tank 37H material represented a challenge in terms of ^{90}Sr removal. The MST used in this demonstration consisted of a batch of material prepared more than 5 years ago, which exhibited a strontium removal capacity well below that specified at the time of purchase. The SRTC chose to use the material, accepting the risk that ^{90}Sr removal may be poor due to the limited quantity of material available at the time of the test and the need to produce material for the subsequent CSSX demonstration for ^{137}Cs separation.

The next demonstration consisted of treating about 30 liters of a composite of Tank 37H and Tank 44F wastes in a reactor scaled to provide the same mixing conditions that will occur in the batch contact tank. The SRTC diluted the composite waste to 5.6 M in sodium ion and 0.4 g/L of MST to affect actinide and strontium removal. The MST used in the test was from a different batch of material and exhibited a strontium removal capacity well-above the minimum specified for the process. Strontium and plutonium decontamination factors measured 18 and 6.2, respectively, which are in good agreement with previous measurements using simulated solutions. Identical decontamination factors were measured in 0.1 L-scale tests, which were well-mixed. Thus, it was concluded that the planned agitation for the batch contact tank is adequate to meet process removal requirements.

The SRTC also performed small-scale (0.1 L) laboratory tests with a multiple-tank composite of Tanks 39H, 33F, 26F, and 21H. These tanks were selected to provide a waste solution that would represent a bounding case for alpha removal. The ^{90}Sr, plutonium, and ^{237}Np activities of the waste solution (diluted to 5.6 M in Na^+) measured 415, 490, and 0.059 nCi/g, respectively (see Table 7). To meet the Saltstone WAC, decontamination factors must exceed 10 for ^{90}Sr; 27 for plutonium; and 2.0 for ^{237}Np. Due to the high plutonium content, SRTC performed three separate strikes of 0.4 g/L MST. The second and third additions of MST occurred 30 hours and 54 hours after the first addition.

Table 7. Initial Sorbate Concentrations and Resulting Decontamination Factors in Tests with Tank Waste with High Plutonium Concentration (Fink et al. 2002b).

Sorbate	Initial Conc. (nCi/g)	Decontamination Factor		
		1st Strike	2nd Strike	3rd Strike
Sr	415	10 (N)	4.0 (Y)	nd (Y)
Pu	490	6.2 (N)	2.9 (N)	1.6 (Y)
Np	0.059	3.8 (Y)	nd (Y)	nd (Y)
U	0.0039	1.2 (Y)	1.4 (Y)	1.2 (Y)

Letters in parenthesis refer to whether the treated solution met the Saltstone Waste Acceptance Criteria, Y = Yes, N = No

Table 7 provides a summary of the measured decontamination factors for the demonstration. Strontium removal was just above the Saltstone WAC after the first strike and well-below the WAC after the second strike. The ^{90}Sr activity fell below detection limits after the third strike and, thus, the SRTC was unable to determine a decontamination factor for the third strike. In contrast to the earlier large-scale demonstration with the Tank 37H waste solution, this batch of MST performed very well in removing strontium.

Figure 12 provides a plot of the total plutonium activity versus time. Plutonium removal occurred with each MST strike. The plutonium activity after the first MST strike measured about 78 nCi/g, which is well-above the Saltstone WAC for alpha activity (18 nCi/g). After the second strike, the plutonium activity decreased to about 24 nCi/g, which is just above the Saltstone WAC. Sufficient plutonium removal occurred after the third strike to meet the Saltstone WAC for alpha activity. Overall, a combined average plutonium decontamination factor of 29 was obtained after three strikes of MST.

Neptunium and uranium removal also occurred. Sufficient neptunium removal occurred from the first strike so that the waste solution met the

Fig. 12. Plutonium Concentration versus Time in Tests with Tank Waste Containing High Plutonium Concentration (Fink et al. 2002b).

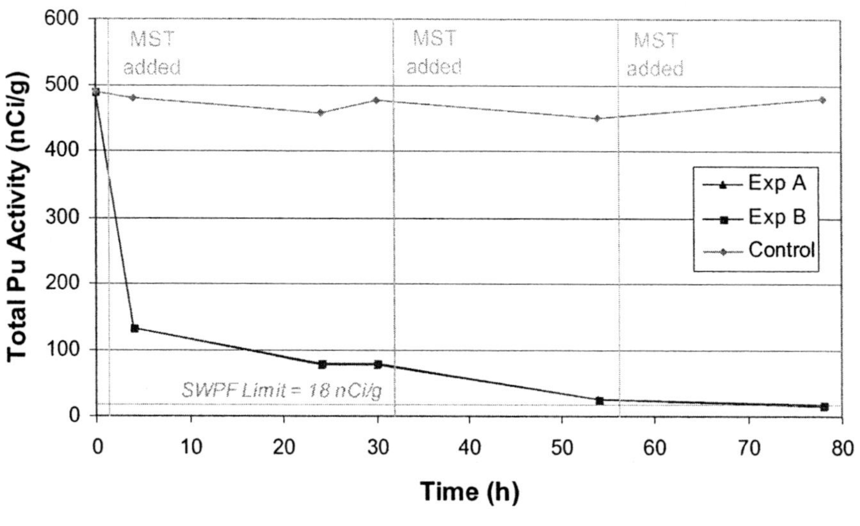

Saltstone WAC for ^{237}Np activity (0.03 nCi/g). The concentration of ^{237}Np fell below the detection limit of the analytical method after the second strike and, thus, the SRTC was unable to determine decontamination factors for the second and third strikes. Uranium removal occurred with each MST strike. The measured decontamination factors agreed well with those measured previously with simulated waste solutions.

Alternative Actinide and Strontium Removal Sorbents

In the past, the HLW program has relied exclusively on process options that use MST to achieve the required removal of actinide and strontium (Harmon et al. 2001a). The DOE and NRC judged such reliance upon MST as the sole technology as an unacceptable technical risk. For example, use of alternate sorbents or technologies open the potential of alternate engineered designs, perhaps using existing equipment, to achieve the required decontamination.

During FY00, the Project Team conducted a review of available literature for data related to a number of actinide and strontium removal technologies. This evaluation recommended the following sorbent materials for further testing to determine the rate and extent of removal: sodium nonatitanate in the form under development by Honeywell Performance Polymers and Chemicals (Morristown, NJ); SrTreat® produced by Selion OY (Finland); CST in various forms; and pharmacosiderites as developed by Clearfield at Texas A&M. The report also recommended evaluating precipitation with $Sr^{2+}/Ca^{2+}/NaMnO_4$. The study recommended not pursuing any testing of liquid/liquid extraction and polymer filtration methods.

A review of the use of sodium nonatitanate began in FY00. Researchers evaluated the performance of three samples of the material relative to actinide and strontium removal. Also, the same samples were evaluated to determine the influence on cross-flow filter performance.

Physical and chemical characterization indicated that the three samples exhibited similar particle volume distributions, which proved larger than that measured for the reference MST material. In actinide and strontium removal testing, the samples exhibited lower removal capacities than MST. Removal rates appeared similar after 24 hours. Review of the x-ray analyses for the sodium nonatitanate suggests that the Honeywell samples represent a poor conversion of the sorbent to the desired structure and appear atypical of the material that the Honeywell production should yield. Based on these data, SRTC recommended additional testing to measure removal kinetics during the first eight hours of contact between the solution and sorbent. It was also recommended that further testing of sodium nonatitanate samples proceed only upon documented evidence that future samples exhibit the structure expected for the synthesized sorbent as determined by x-ray diffraction.

Bench-scale dead-end filtration tests used 5.6 M sodium, average salt solution containing 0.6 g/L simulated sludge, and 0.55 g/L MST or sodium nonatitanate. Testing identified no correlation between MST or sodium nonatitanate particle size and filter flux. Any potential filtration gains from differences in particle size between the MST and sodium nonatitanate appeared offset by changes in filter cake porosity. The dispersion of the particle size for these samples likely contributes to this behavior. The sodium nonatitanate particles produced marginal improvement in filter flux (~30%). The rate of improvement in filter flux proves less than previous gains obtained through the addition of chemical additives to improve performance. The marginal improvement would not appreciably reduce equipment size.

The SRTC conducted screening tests of actinide and strontium removal with crystalline silicotitanate (CST), SrTreat®, and sodium nonatitanate samples; and by precipitation upon addition of nonradioactive strontium, calcium, and sodium permanganate. Several of the alternate sorbents and the precipitation method exhibited actinide and strontium removal characteristics as good as or better than the reference MST material (Fig. 13). Consequently, it was found reasonable to continue evaluation of the sodium nonatitanate and SrTreat® materials and the permanganate precipitation method as alternatives to MST for a batch process to remove strontium and actinides. Such efforts focused on improving plutonium and neptunium removal characteristics to ensure that the Saltstone limits can be met at bounding waste conditions.

Currently, evaluations of new sorbents are continuing. However, at present, the Project Team has not identified a commercially-available sorbent that is superior to MST.

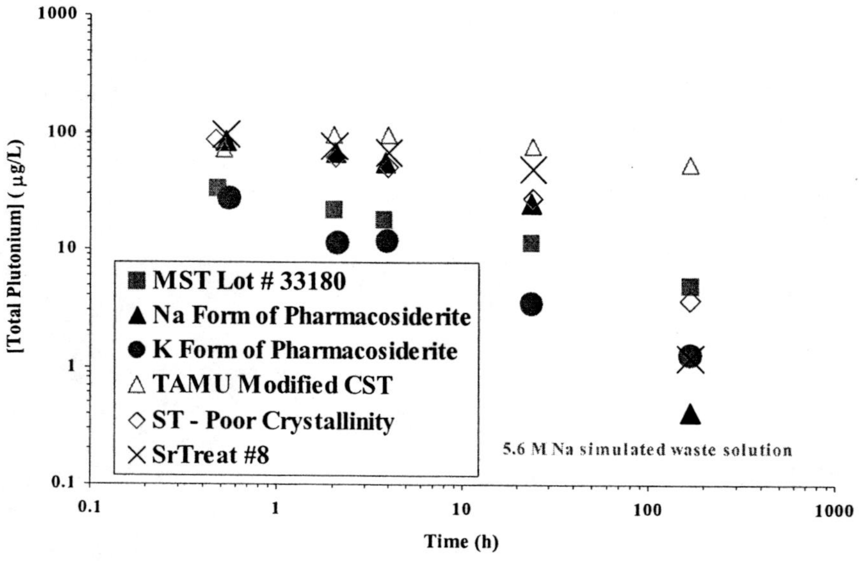

Fig. 13. Plutonium Removal with MST, ST, CST, and Pharamacosiderate Materials (Barnes et al. 2001).

MST/Sludge Cross-Flow Filter Testing

Cross-Flow Filter Testing (0.5 µm Pore Size): Tests on filtration of MST/sludge simulant mixtures alone, with additives and flocculants to improve filtration rates, were performed during Phase IV. Various MST/Sludge simulant mixtures were filtered up to ~5 wt % in the SRTC Parallel Rheology Experimental Filter (PREF). The filter fluxes were, on average, about 0.02 gallons per min/ft^2 (0.84 L min^{-1} m^{-2}) and did not vary significantly with composition (Fig. 14). Bentonite and polyethylene oxide, the most promising additives tested, increased filtration rates by 25-35%. Hydroxamated amines by Cytec and Alclar® 600, the most promising flocculants tested, increased filtration rates by 25-30%.

Engineering-scale cross-flow filter tests were conducted at the University of South Carolina using the Filtration Research Engineering Demonstration (FRED). The testing examined operating conditions that could improve throughput and reduce the filter size needed for solid-liquid separation in the actinide and strontium removal step of the Salt Waste Processing Facility (SWPF). The filter test used 5.6 M sodium, average salt solution

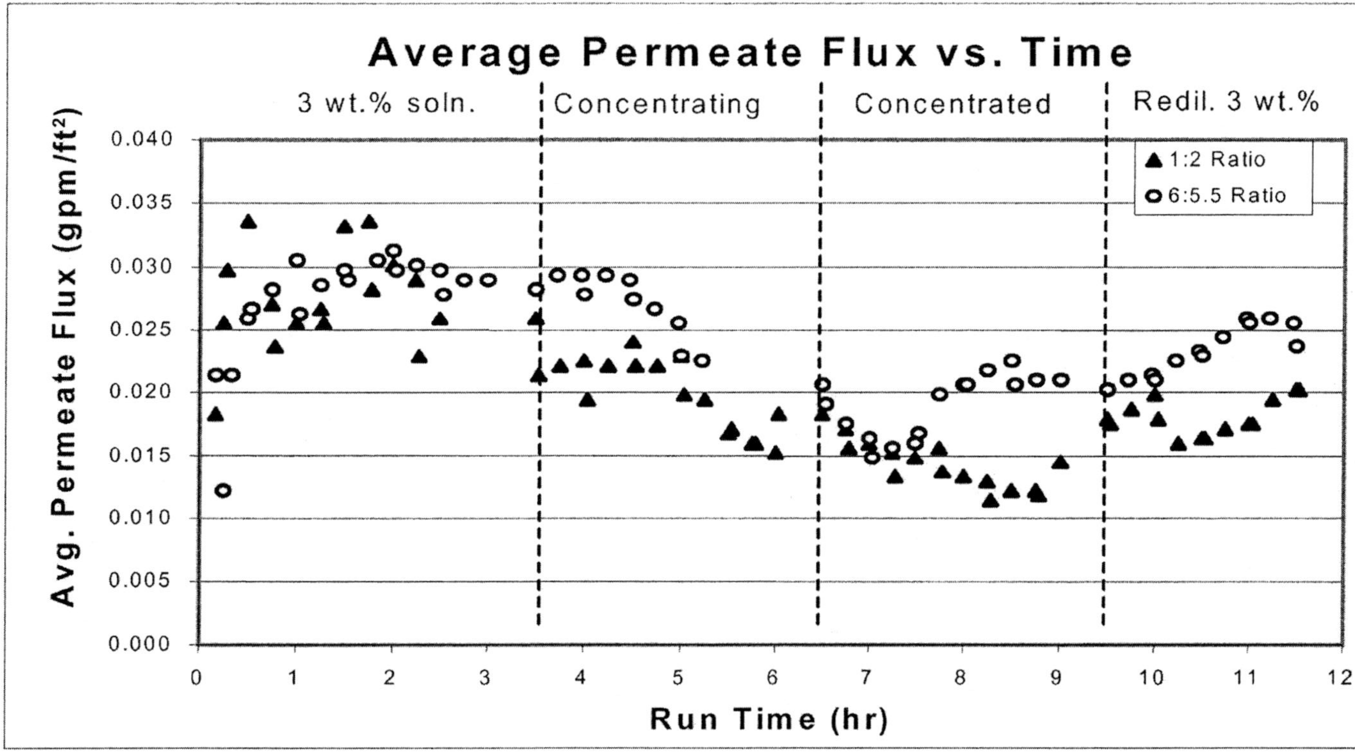

Fig. 14. MST/Sludge Filtration Rates (Dimenna et al. 2001).

with simulated sludge (Tank 40H and Tank 8F) and MST added. The insoluble solids concentration varied between 0.331 and 41.9 g/L. The transmembrane pressure (TMP) varied between 12 and 68 psi (83 kPa and 469 kPa). The axial velocity varied between 12 and 26 ft/s (3.6 and 7.8 m/s).

The significant conclusions of the testing follow:

1. Filter fluxes greater than 0.05 gpm/ft^2 (2.10 L min^{-1} m^{-2}) occurred at all insoluble solids concentrations tested.
2. Filter flux proved most sensitive to TMP. Filter fluxes as high as 0.14 gpm/ft^2 (5.88 L min^{-1} m^{-2}) resulted with a TMP of 68 psi (469 kPa).
3. The filter flux shows a strong correlation with increasing concentration causing a decrease in filter flux. The filter flux decreased significantly as the concentration of solids exceeded 1 wt % (Fig. 15).
4. The filter flux data shows a dependence on axial velocity, but this dependence proves an order of magnitude less than the dependence on TMP and concentration.
5. Backpulsing and scouring can effectively recover filter performance under the conditions examined.

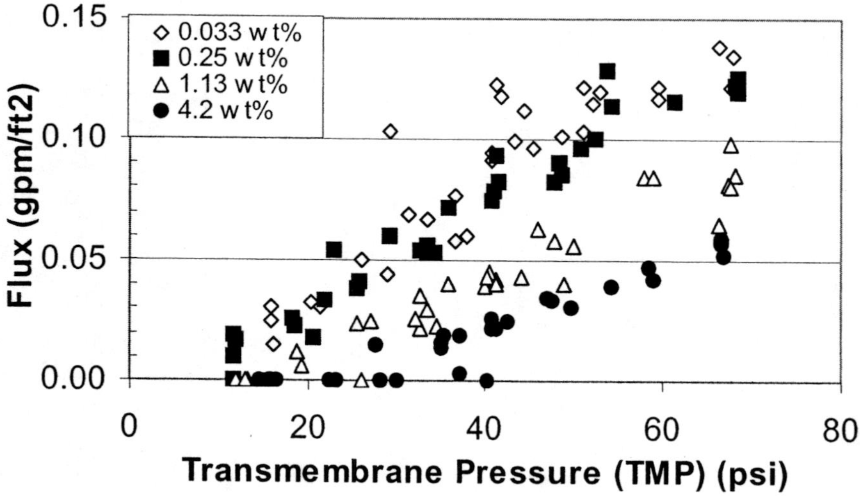

Fig. 15. Pilot-scale Cross-flow Filter Testing (Poirier et al. 2002).

Actual waste filtration test data from testing using the Savannah River Technology Center (SRTC) Cells Unit Filter (CUF)—with actual sludge samples combined with MST and supernate from HLW Tanks 37H and 44F—show reasonable agreement between filtration with actual waste and filtration testing with simulant (Fig. 16). The filtration testing with actual waste confirms that simulant testing provided a conservative design basis for filtration equipment sizing.

Cross-Flow Filter Testing (0.1 μm Pore Size): The prior efforts on the conceptual design of the SWPF assumed use of a 0.5 μm pore-size, cross-flow filter (Poirier and Fink 2002). Use of this configuration was derived from prior experience at SRS using the equipment for filtration of slurries containing sludge, MST, and tetraphenylborate solids. The Project Team examined the use of this filter media in several demonstrations with simulated and actual wastes in equipment at laboratory-scale through pilot scale. However, the SRTC observed passage of solids through one such filter at the University of South Carolina, raising doubts as to the adequacy of the selection. Similarly, the waste treatment program at Hanford adopted the use of 0.1 μm elements for the analogous step in the process based in large part on proprietary studies conducted by British Nuclear Fuels Limited in support of the Sellafield operations.

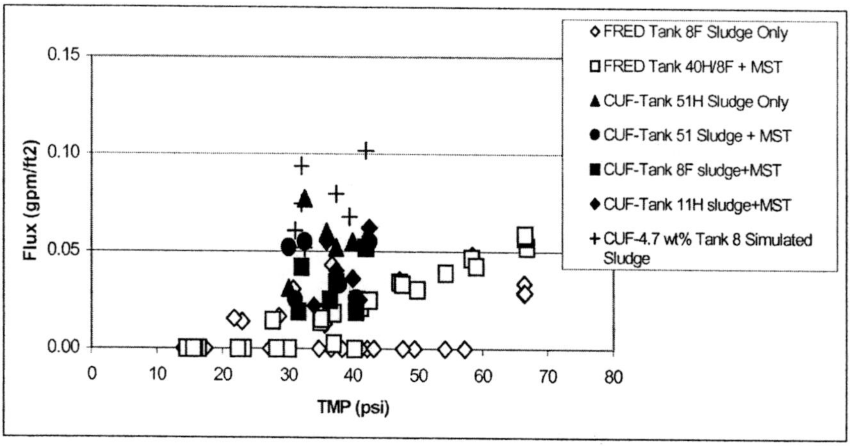

Fig. 16. CUF Real Waste MST/Sludge Filtration Data at 6 wt% Solids (Dimenna et al. 2001).

In light of these observations, the Project Team elected to examine the use of 0.1 μm elements to judge their relative performance. The SRTC conducted a series of laboratory tests using both simulated and actual wastes. Figure 17 shows similar filter performance from experiments using actual waste. Again, these studies used operating pressures and velocities expected to bound those anticipated in the SWPF. The sample of sludge slurry came from Tank 40H. The SRTC treated the sludge slurry with either MST or sodium permanganate to remove soluble strontium and alpha-emitting actinides. Testing used a single filter element. For either process option, the 0.1 μm element provided equal or superior performance. Based on experimental studies with simulated and actual wastes, SRTC recommend adopting the use of 0.1 μm pore-size, sintered stainless-steel filter elements within the design of the SWPF.

Given the limited duration of the laboratory testing, longer duration studies were conducted with simulated waste at larger scale. Those studies examined the impact of the smaller pore size on frequency of the cleaning operation, and on the processing rate for the larger-scale equipment. Testing at the University of South Carolina supported this program recommendation.

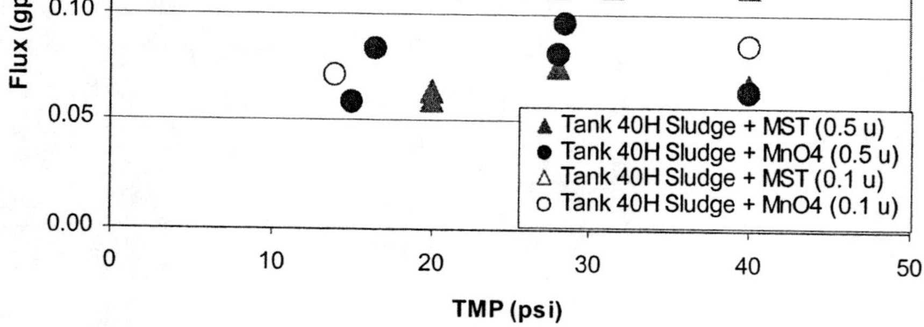

Fig. 17. Relative Filter Performance from Experiments Using Actual Waste Slurries and Filter Elements with Different Pore Size (Poirier and Fink 2002).

Alternative Solid/Liquid Separations

The SRTC conducted an evaluation of alternate methods for achieving the required separation of solids from liquid. Based on the findings, SRTC recommended evaluation of several alternate solid-liquid separation technologies for removing sludge and MST from HLW salt solutions. In continuing work in this area, primary focus was directed toward identification of chemical additives (e.g., flocculating agents) that will improve the performance of the cross-flow filters. Other work should investigate settling and decanting followed by polishing filtration (both cross-flow and dead-end). If flocculation with cross-flow filtration proves ineffective, SRTC planned to investigate high-shear filtration (using a centrifugal or Vibrator Shear-Enhanced Processing filter) as well as flocculation in combination with centrifugation.

The SRTC conducted measurements of the influence of various chemical additives on solid-liquid separation using simulated waste. However, the additives yielded only marginal improvement in flux when using a cross-flow filter. It was speculated that the reduced performance results were due to shearing of the aggregates at the hydraulic conditions of the filtration. Based on these findings, the Project Team deferred future testing at pilot-scale and with actual waste until completion of studies examining the use of the additives in a process configuration that incorporates sedimentation and decanting prior to filtration.

The SRTC coordinated a set of vendor tests examining a centrifugal, high-shear implementation of cross-flow filtration (rotary microfilter) for simulated waste containing metal hydroxide sludge and MST. The filtrate throughput proved higher than the facility design target, and reached as much as about four times observed in comparable testing using the pilot-scale equipment for conventional cross-flow filtration. The majority of the testing used a composite filter media of ceramic and stainless steel rated as 0.1 μm in pore size. Additional testing, albeit more limited in scope, showed nearly identical performance using a 0.5 μm stainless steel filter.

The SRTC conducted additional filtration tests using a SpinTek Model II-1 rotary microfilter using both simulated and actual waste. The actual waste feed for the tests consisted of SRS HLW supernate from Tank 37H; SRS

HLW sludge from Tank 51H; strontium nitrate; a sodium permanganate; and MST. One slurry contained MST and sludge to achieve calculated insoluble solid concentrations of approximately 0.09, 1.29, and 4.5 wt %. A second slurry contained sufficient sodium permanganate to achieve calculated insoluble solids concentrations of 0.23, 3.28, and 11.7 wt %. Testing used a filter disk with nominal pore size of either 0.1 μm or 0.5 μm.

Data with simulant and MST agreed well with previous testing of the rotary microfilter using simulant showing a 2 to 6 times increase in filtration rate when using a 0.1 or 0.5 μm rotary filter in place of a 0.5 μm cross-flow filter.

Results for actual waste testing with MST and a 0.1 μm rotary filter showed 0 to 100% improvement in filter flux compared to prior tests using a 0.5 μm cross-flow filter. Filtration rates using permanganate showed little improvement in filtration rate using the 0.1 μm rotary filter as compared to the 0.5 μm Mott cross-flow filter. Analysis of the data as a function of time and post-test investigation suggests the 0.1 μm filter disk deteriorated over the course of the experiments. The SRTC believed the flow path created by the permeate carrier became successively constricted during the experiments causing a continual decline in performance. Hence, these experiments resulted in a low bias for the performance of the equipment.

The SRTC modified the design of the permeate carrier for the 0.5 μm rotary filter. Testing with the 0.5 μm rotary filter gave an increase in filtration rate of 3 to 10 times over the 0.5 μm cross-flow filter for feed slurries treated with both MST and sodium permanganate (Fig. 18). A significantly higher flux rate occurred for the 0.5 μm rotary filter versus the 0.1 μm rotary filter. Additional equipment modifications for radioactive service are under evaluation.

MST/Sludge Resuspension

Tests were performed at ORNL to study the rheology and resuspension characteristics of MST/sludge slurries at both laboratory-scale and pilot-scale. Tests showed that settled slurry is relatively easy to resuspend at pilot-scale after settling for 14 days—although it is possible that not all the monosodium titanate (MST) was resuspended.

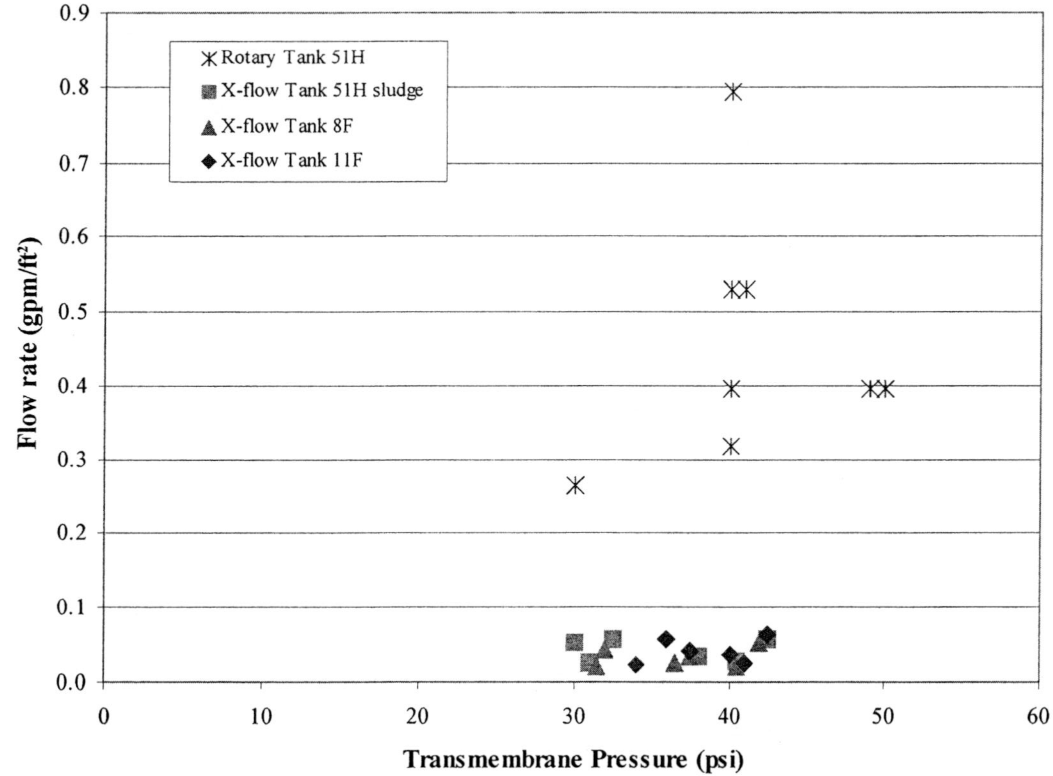

Fig. 18. 0.5-Micron Rotary Filter and 0.5-Micron Cross-Flow Filter with 4.5 wt % Insoluble Solids Using Actual Waste and MST (Herman et al. 2003).

After 60 days of settling time, the slurry could not be completely resuspended even at an impeller tip-speed of 300 m/min. Storage of MST/sludge mixtures at 80°C (for as little as 3 days) had a dramatic effect on yield stress and consistency. After 60 days at 80°C the yield stress increased by a factor of 300, and the consistency by a factor of 30. These measurements indicate the MST/sludge must be cooled to assure resuspension, in the event agitation is lost for an extended period. The Alpha Sorptium Tank (AST), the Sludge Slurry Receipt Tank (SSRT), and the downstream tanks (intermediate pump pit tanks and MST/Sludge processing tanks located in the DWPF) will require both cooling coils and high-powered/high tip-speed agitators to ensure resuspension of settled solids.

Additional studies were conducted in late FY-03 and FY-04 to determine if the resuspension concerns will occur at lower temperatures. Final reports are in preparation.

CAUSTIC-SIDE SOLVENT EXTRACTION

Process Description

In solvent extraction, a sparingly soluble diluent material containing an extractant to complex the cesium ions is mixed with the aqueous caustic solution to remove cesium. The decontaminated aqueous stream (raffinate) is then sent to the Saltstone Production Facility (SPF) for disposal. The cesium contained in organic solution is then stripped into an aqueous phase ready for transfer to the Defense Waste Processing Facility (DWPF). The solvent is cleaned to remove impurities and recycled. Prior to treatment by solvent extraction, actinides and strontium are removed from the waste by sorption with the MST. The resulting slurry is then filtered to remove the MST and sludge solids. The CSSX process uses a novel solvent system (Fig. 19) containing the following four components (Harmon et al. 2002a):

1. An extractant, calix[4]arene-bis(tert-octylbenzo-crown-6), also called BOBCalixC6, which is a calixarene crown specific for cesium extraction
2. A modifier, 1-(2,2,3,3, -tetrafluoropropoxy)-3-(4-sec-butylphenoxy)-2-propanol, also called Cs-7SB, which is an alkyl aryl polyether that keeps the extractant dissolved in the solvent and increases its ability to extract cesium in the extraction section
3. A suppressant, tri-*n*-octylamine (TOA), which suppresses effects from organic impurities to ensure that the cesium can be back-extracted from the solvent in the strip section
4. A diluent, Isopar® L, which is a mixture of branched hydrocarbons

The solvent is contacted with the alkaline waste stream in a series of countercurrent centrifugal contactors (the extraction stages) where cesium and nitrate are extracted into the solvent phase. The resulting clean aqueous raffinate is transferred to the SPF for conversion to saltstone. Following cesium extraction, the solvent is scrubbed with dilute acid to remove other soluble salts—particularly sodium and phosphates from the solvent stream (the scrub stages). The scrubbed solvent then passes into the strip stages where it is contacted with a very dilute acid stream to transfer the cesium to the aqueous phase. The aqueous strip effluent

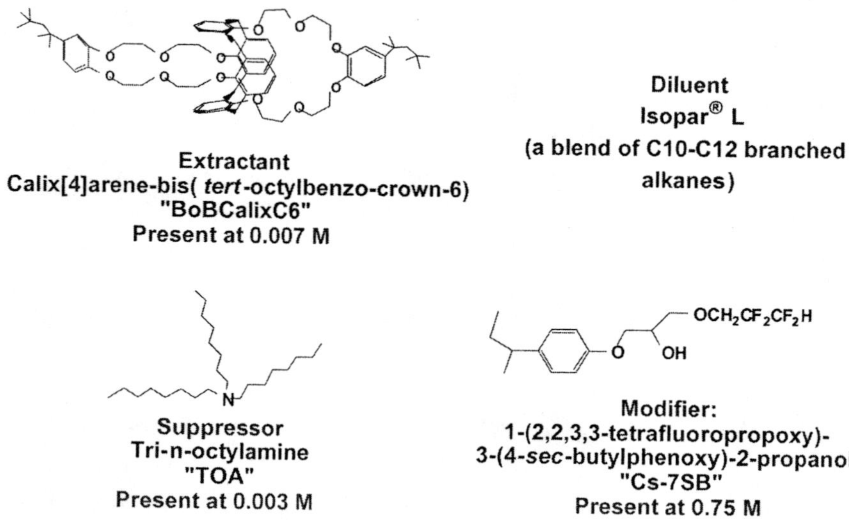

Fig. 19. Solvent Components (Moyer et al. 2002).

containing pure cesium nitrate, which is 15 times more concentrated than in the waste, is transferred to the DWPF for vitrification. Figure 20 (Dimenna et al. 2001) contains a schematic representation of the proposed solvent extraction flowsheet.

In the strip stages, the presence of lipophilic anionic impurities (e.g., dibutylphosphate, dodecylsulfate) has the potential to greatly reduce stripping performance. Such impurities could possibly come from the waste or from solvent radiolysis. To remedy the potential effects of these impurities, tri-*n*-octylamine (TOA) is added to the solvent. This amine remains essentially inert in the extraction section of the process but converts to the trioctylammonium nitrate salt during scrubbing and stripping. This salt remains in the organic phase and allows the final traces of cesium in the solvent to be stripped by supplying any anionic impurities in the solvent with equivalent cationic charges.

Over long periods of time, either the modifier, the TOA, or the calixarene may degrade chemically or radiolytically. The most likely degradation is that of the modifier to form a phenolic compound that is soluble in the organic phase in contact with acid solutions. However, the modifier was

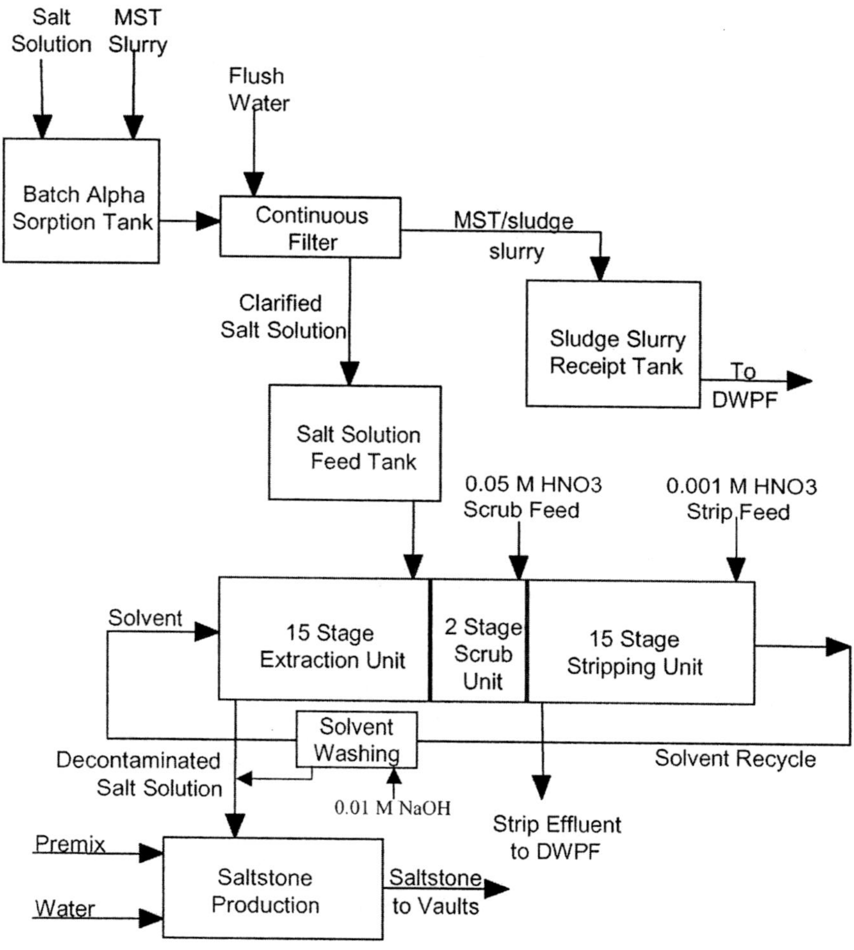

Fig. 20. Schematic Representation of CSSX Process (Dimenna et al. 2001).

designed so that the phenolic compounds would distribute preferentially to alkaline aqueous solutions—either the waste itself or NaOH wash solutions. Gradual degradation of the solvent will result in some loss of performance, owing both to the loss of calixarene, modifier, and amine and to the buildup of various degradation products. The proposed flowsheet contains two work stages: Stage 1) an acidic wash of the solvent; and Stage 2) a caustic wash of the solvent to maintain solvent performance. These two wash stages are intended to remove any acidic or caustic impurities that may accumulate in the solvent system over time. In particular, the caustic wash is known to remove the modifier degradation products. In addition, the proposed flowsheet has assumed that the solvent will be replaced on an annual basis to maintain system performance. Spent solvent will be incinerated.

The aqueous output streams from the CSSX process may contain either soluble solvent components and/or entrained organic phase. This potential loss may represent an economic concern due to the expensive solvent components or a problem in downstream operations. The proposed process contains solvent recovery processes for the aqueous effluent streams. Additional contactor stages are provided to remove soluble organic compounds, and in particular, to remove solvent from the exiting streams with a small amount of Isopar® L. The aqueous phase from these stages is then sent to a settling tank where any remaining entrained organic (mostly the Isopar® L) is allowed to float and is decanted. The Isopar® L containing solvent is distilled to recover the extractant and modifier. The Isopar® L added in the two solvent recovery processes is sent to the Consolidated Incinerator Facility.

Strip effluent storage is provided to accommodate the differences in cycle times for the Slurry Receipt Adjustment Tank (SRAT) in Defense Waste Processing Facility (DWPF), and to allow for disengagement of any organic carry-over from the extraction process. Strip effluent will eliminate the need for an evaporator. The strip effluent is evaporated in the DWPF SRAT where the nitric acid content is used to offset the nominal nitric acid requirement. The effluent would contain <0.01 M sodium, and <0.001 M of other metals.

Caustic-Side Solvent Extraction Chemical and Physical Properties

All development and testing of the CSSX process through 2001 used a baseline solvent composition of 0.01 M BOBCalixC6, 0.50 M Cs-7SB, and 0.001 M TOA in Isopar® L (Harmon et al. 2002a). In 2002, the composition of the CSSX solvent was modified to 0.007 M BOBCalixC6, 0.75 M Cs-7SB, and 0.003 M TOA in Isopar® L (Klatt et al. 2002). This change was required to address BOBCalixC6 saturation and solvent stability issues. All tests in 2002 were performed using the optimized solvent composition.

Technical risks of the CSSX process (Harmon et al. 2002a) that require specific attention include the following:

1. Chemical stability of the solvent matrix
2. Radiolytic stability of the solvent matrix
3. Proof-of-concept performance of the proposed process flowsheet with simulated waste
4. Performance of the CSSX flowsheet with actual SRS high-level waste

Under a wide range of expected operating conditions but within the expected operating temperature range 20-35°C, Moyer et al. (2002) reported that no issues were identified regarding extraction, scrubbing, and stripping (ESS) behavior (Fig. 21). Information on cesium distribution in ESS steps serves as input for flowsheet design; provides a baseline for evaluating solvent performance under varying conditions; and contributes to an understanding of the effects of expected process variables. In parallel, other measurements were directed toward learning how other system components distribute in the flowsheet, including the solvent components, waste constituents, and solvent degradation products. Upon understanding which components influence flowsheet performance, it was then possible to address cleanup of the solvent to maintain its stable function. The waste simulant was a complex mixture defined by SRS. A significant temperature dependence that causes the cesium distribution ratio (D_{Cs}) to decrease 6-10% per degree with increasing temperature necessitates process-temperature management. This property can be turned to an advantage by exploiting dual temperature control of extraction and stripping. The chief competing species, K^+, has the potential to depress Cs^+ extraction, but only when K^+ exceeds its bounding concentration in the waste (ca. 50 mM). Extraction of Na^+ also occurs but does not compete significantly. Extracted K^+ and Na^+ partition almost completely to the scrub solution, as do bulk inorganic anions such as nitrite and chloride. Trials of various simulant recipes demonstrated the robustness of the solvent to handle widely-varying bulk waste compositions. Elements present in the simulant, including aluminum, chromium, copper, iron, molybdenum, mercury, silver, lead, palladium, rhodium, ruthenium, zinc, tin, and fluorine, could not be detected in the solvent after five cycles, nor do they appear in the scrub or strip stages in significant concentrations. Five trace metals added to the full simulant recipe, (namely uranium, neptunium, plutonium, strontium, and technetium) were not significantly extracted, nor did they affect cesium extraction or stripping.

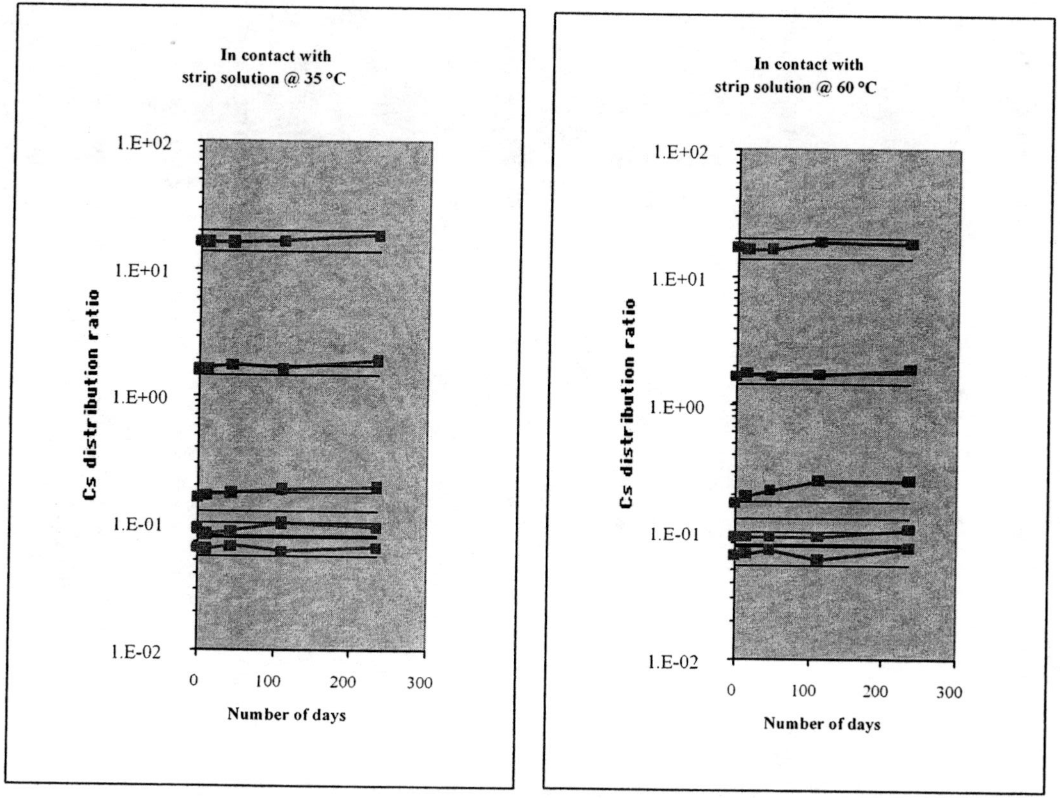

Fig. 21. ESS Experiments (Moyer et al. 2002).

The aqueous pH of scrub and strip solutions measured after contacting reflects the expected acid uptake by the TOA in scrubbing and partial release in stripping (Moyer et al. 2002). Good phase separation is required, as excessive entrainment of the simulant in the solvent can neutralize the low acid scrub and strip solutions resulting in poor stripping of cesium from the solvent. A survey of the effect of minor organic species confirmed that lipophilic anions such as dibutylphosphate, traces of which are present in the waste, can build-up in the solvent on multiple cycles and impair stripping performance. These anions—together with 4-sec-butylphenol (a breakdown product of Cs-7SB) and surfactants having 12 or less carbon atoms per molecule—can be readily removed by washing the solvent with dilute NaOH solutions. Passing the solvent through a strong-base anion exchange resin in the hydroxide form was effective for removal of an 18-carbon surfactant. The neutral compound tributylphosphate and 1-butanol partition significantly to the solvent, but were not found to affect performance for the expected levels in the waste. Dioctylamine, a degradation product of TOA, is expected to partition to the strip effluent. Trimethylamine, present in trace quantities in the waste, can potentially build-up in the solvent. Although it was not associated with impaired performance, its behavior and effects remain to be understood.

Moyer et al. (2002) reported that the solvent exhibited high thermal stability in batch tests. Acceptable ESS performance was sustained for 235 days of constant exposure of the solvent to the process solutions at 35°C. The only solvent component that displays any sign of chemical instability is the TOA, as manifested under severe off-normal conditions involving 110 days or longer of continuous contact with the scrub solution at 61°C. The resulting decrease in TOA concentration correlates with an increase in the cesium distribution ratios in the scrub and stripping steps. Normal performance was re-established upon replenishing the TOA and solvent washing. The chief degradation product of TOA was shown to be dioctylamine, which is expected to be washed-out in the aqueous strip effluent. Potential catalysts in the full simulant (e.g., copper, iron, molybdenum, mercury, silver, lead, palladium, rhodium, and ruthenium) have no apparent ability to accelerate degradation. Nitration reactions were not detected under process conditions. The only nitration detected was that of the modifier upon heating of the solvent for 28 days at 61°C with 1-M nitric acid.

Based on the data obtained from solvent samples externally irradiated at the SRS, no issues were identified regarding the radiation stability of the solvent. Neither significant degradative loss of the primary solvent components nor consequential impairment of ESS performance was observed over dose ranges well in excess of annual levels. For BOBCalixC6 and Cs-7SB, loss due to radiolytic breakdown was estimated to require less than 1% of annual makeup. For TOA, the annual loss is expected to be 10%. The primary degradation product observed was 4-sec-butylphenol, due to a slight breakdown of the modifier. Traces of organic products, especially fluorinated products from the modifier, distribute to the aqueous phases. Interfacial behavior was also acceptable, as measured by break times and interfacial tension. In batch tests, extraction and scrubbing behavior remained acceptable. Stripping was also acceptable for all but the 16 Mrad dose, where poor performance was related to the loss of 79% of the TOA. Except for the most deteriorated solvent samples, replenishing lost TOA and solvent washing with NaOH solutions restored performance.

Examination of solubility phenomena raised several concerns (Harmon et al. 2002a). Although no issues were identified for the solubility of the baseline modifier Cs-7SB, the alternative modifier Cs-6 formed an insoluble hydrate. The baseline solvent was supersaturated in BOBCalixC6 at 25°C. Precipitation and dissolution of BOBCalixC6 are apparently very slow, however, and observation of the baseline solvent under various conditions for 224 days revealed no tendency to form solid phases. Considering that BOBCalixC6 solubility increases to acceptable levels as the Cs-7SB concentration increases, it was recommended that a new solvent composition with higher Cs-7SB concentration be adopted. Third-phase formation was found to occur as a result of high Cs^+ or K^+ loading. Although high Cs^+ loading is not expected to be found in the flowsheet, K^+ loading is expected to be significant. If the temperature falls below 20°C, the solvent tolerance for K^+ is exceeded at the bounding aqueous concentration, setting an effective lower operating limit of the process. However, raising the Cs-7SB concentration could lower this temperature limit.

In the absence of solvent recovery, minor solvent replacement will likely be required owing purely to solubility losses to the aqueous phase. Partition ratios for BOBCalixC6, Cs-7SB, and TOA in the baseline solvent in contact with the process aqueous phases at 25°C are high, exceeding the analytical limits of measurement. Conservatively, the analytical limit for BOBCalixC6 implies an acceptable maximum cumulative loss of 49% of the BOBCalixC6 annually. The primary need for solvent-washing arises from the necessity of removing extractable minor waste components such as dibutylphosphate. The solvent breakdown products (4-secbutylphenol and dioctylamine) both partition respectively into the simulant and strip aqueous phases, rendering the solvent effectively free of all but traces of these compounds, even without solvent-washing. A solution containing 10-mM NaOH was recommended as a suitable wash solution for the extended contactor tests and for further flowsheet evaluation. Tests of solvent-washing with various concentrations of NaOH, including 10-mM NaOH, were generally successful, allowing restoration of solvent function after multiple cycles with the full simulant. It also restored function to the thermal-stability and internal-irradiation test solvents after replenishing the TOA that had been degraded. It did not, however, restore function on archived samples from the external-irradiation test at the SRS.

The vapor pressure of the 1-(2,2,3,3-tetrafluoropropoxy)-3-(4-sec-butylphenoxy)-2-propanol was measured over the temperature range 62°C to 92°C using a Knudsen effusion technique (Steele 2002). The vapor pressure is related to temperature by the following equation:

$$\ln(p/Pa) = (32.202 \pm 0.265) - (12154 \pm 93)/T$$

where p is the pressure (expressed in Pa), Pa is the reference pressure of 1 Pa, and T is the temperature (expressed in K). The derived heat of vaporization is (101.1 ± 0.8) kJ/mol at 351 K. Because the vapor pressures over the temperature range from 15°C to 50°C were lower than the design capabilities of the Knudsen effusion apparatus, the vapor pressures at these temperature limits were obtained by extrapolation. These estimated values are $(4.6 \pm 0.3) \times 10^{-5}$ Pa for 15°C, and $(4.5 \pm 0.1) \times 10^{-3}$ Pa for 50°C.

Data for the solubility of the extractant, calix[4]arene-bis(tert-octylbenzo-crown-6), showed the baseline solvent composition to be supersaturated with respect to the extractant. Although solvent samples have been observed for approximately one year without any solids formation, work was completed with the following objectives:

1. To define a new solvent composition that is thermodynamically stable with respect to solids formation.
2. To expand the operating temperature with respect to third-phase formation.

The following data were mapped against a set of predefined performance criteria: calix[4]arene-bis(tert-octyl benzo-crown-6) solubility; cesium distribution ratio under ESS conditions; flowsheet robustness; temperature range of third-phase formation; dispersion numbers for the solvent against waste simulant, scrub and strip acids, and NaOH wash solutions; solvent density; viscosity; and surface and interfacial tension. The optimized composition of 0.007 M calix[4]arene-bis(tert-octylbenzo-crown-6), 0.75 M 1-(2,2,3,3-tetrafluoropropoxy)-3-(4-sec-butylphenoxy)-2-propanol, and 0.003 M TOA in the diluent Isopar® L provided the best match between the measured properties and the performance criteria. Therefore, this optimized solvent is recommended as the new baseline solvent composition (Klatt et al. 2002). Although some minor compromises within acceptable limits were made in flowsheet robustness and solvent density, significant benefits from solvent optimization were gained regarding the lower risk of third-phase formation (operating to as low as 11°C at high cesium and potassium loading, versus 20°C using the previous baseline solvent) and lower solvent cost (Delmau et al. 2002a).

The solubility of BOBCalixC6 in Isopar® L containing 0.5-1.0 M Cs-7SB at 15-35°C was examined (Delmau et al. 2002a) by observing the concentration of BOBCalixC6 as it precipitated from supersaturated solutions over a period of approximately 9 months (Fig. 22). Its solubility at 0.75 M Cs-7SB was found to be 7.7 mM at 25°C. Extraction, scrubbing, and stripping tests showed that the optimized solvent performs acceptably; the extraction strength from baseline waste simulant is lower by 20%, but this is compensated partially by 5-14% improved stripping. When the optimized solvent was washed with an aqueous solution of 10 mM NaOH following an ESS test, identical D_{Cs} values were obtained on a second ESS test with fresh aqueous phases. The temperature dependence of D_{Cs} values in ESS testing was found to be similar to that of the former baseline solvent, and parameters are reported for the estimation of D_{Cs} values for ESS contacts in the range of 15-35°C.

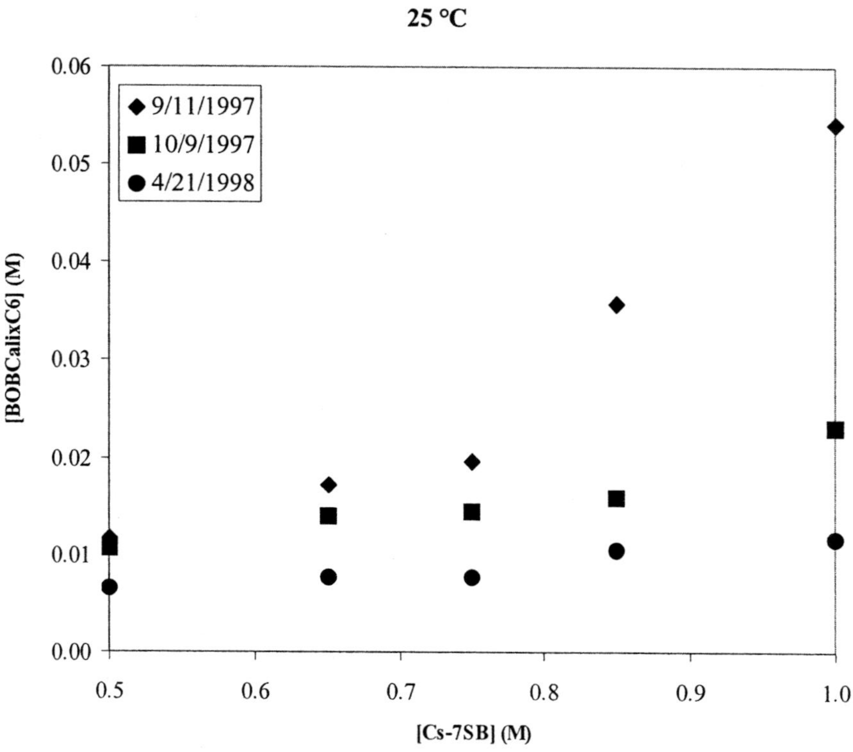

Fig. 22. BOBCalixC6 Solubility Tests at 25°C (Delmau et al. 2002a).

Measurements of the partition ratios of BOBCalixC6, Cs-7SB, and TOA between the optimized solvent and aqueous process solutions indicated that the losses of these solvent components by partitioning to the aqueous process solutions would be minor (Delmau et al. 2002a). Respective fractional replacements of the three components, due solely to partitioning losses over the course of one year, are estimated to be 4%, 27%, and 9%. The partitioning behaviors of selected organic anions examined in studies of the previous baseline solvent were repeated with the optimized solvent, with no change in behavior found. For example, dibutylphosphate, present in the waste in trace amounts, will partition weakly into the solvent on extraction, and the fraction that is extracted will remain in the solvent to be washed-out efficiently by the NaOH wash stage. Similarly, trace metal distribution was not significantly changed. Trace silver, chromium, copper, mercury, manganese, molybdenum, lead, palladium, rhodium, ruthenium, tin, and zinc were not extracted. Traces of aluminum, calcium, strontium, and iron were detected in the optimized solvent after extraction, but were scrubbed-out. The bulk metals sodium and potassium are also extracted as expected, and these are also scrubbed-out. Technetium in the form of pertechnetate anion is extracted very weakly from the simulant (D_{Tc}: 0.038 or 1% extracted), and like dibutylphosphate, it remains in the solvent in the acidic scrub and strip stages—to be washed-out subsequently in the NaOH wash stage.

On storage at or below 0°C, both the previous and optimized solvents exhibit phase splitting in which the more dense of the two phases is concentrated in BOBCalixC6 and Cs-7SB. On warming to room temperature and remixing, the solvents are restored to their original composition and ESS performance. A 43-day test of the solvent stability to elevated temperatures (35°C and 60°C) showed no evidence of degraded ESS performance. The physical properties density, viscosity, and dispersion numbers for the solvent against full simulant, scrub, and strip solutions at 15-35°C were measured and found to be adequate for contactor operation. Dispersion numbers (N_{Di}) under all conditions met the criterion $N_{Di} > 4 \times 10^{-4}$. The solvent density and viscosity, at 25°C, were respectively 0.8516 ± 0.0001 g/m³ and 3.51 ± 0.1 cP (Delmau et al. 2002a).

The CSSX solvent density increases have been observed in separate experimental programs performed by different groups of researchers. Solvent density changes indicate a change in chemical composition. Increased density adversely affects separation of solvent from more dense aqueous solutions present in the CSSX process. Identification and control of factors affecting solvent density are essential for the design and operation of the centrifugal contactors (Lee et al. 2002).

In evaporation experiments, pristine solvent was subjected to evaporation under quiescent conditions at 25°C, 35°C, and 45°C with continuous flowing dry air passing over the surface of the solvent (Fig. 23). Density and viscosity were measured periodically, and chemical analysis was done on the solvent samples. Chemical interaction tests were completed to determine if any chemical reaction took place over an extended contact time that changed the composition or physical properties.

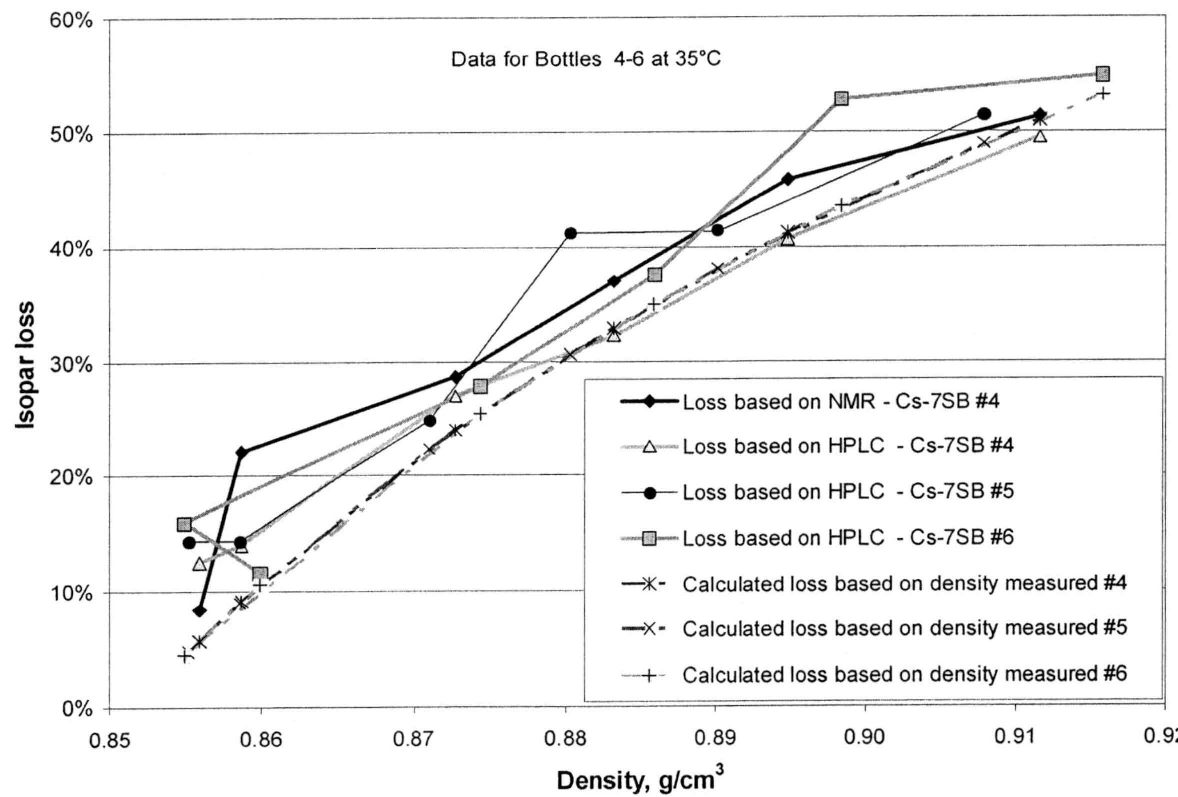

Fig. 23. Solvent Density Versus the Measured and Calculated Diluent Loss at 35°C (Lee et al. 2002).

Solvent and simulant; solvent and strip solution; and solvent and wash solution were contacted continuously in agitated flasks. They were periodically sampled for density (and in some cases viscosity); measurements; and for analysis by nuclear magnetic resonance spectrometry and high performance liquid chromatography using the pristine solvent as the baseline. Chemical interaction tests showed that solvent densities and viscosities did not change appreciably during contact with simulant, strip, or wash solution. There were no effects on density and viscosity, and no chemical changes in the solvent within experimental limits (Lee et al. 2002).

Evaporation test results showed that all solvents were evaporated to densities of >0.90 g/cm^3. Viscosities increased from 3.5 cP to >6.5 cP as the densities increased. Nuclear magnetic resonance spectrometry and high performance liquid chromatography data indicate that: 1) diluent loss is the primary reason for density increase; and 2) the ratio of BOBCalixC6 to Cs-7SB remained almost constant. Density and viscosity vary linearly with the loss of diluent and Cs-7SB concentration increase. Solvent viscosity and density are both sensitive indicators of the loss of diluent, especially when the loss is >10%; density, however, is more reliable at low solvent loss. The ratio of Cs-7SB to BOBCalixC6 appears relatively constant during evaporation to losses of more than 50% of the diluent. A simple density model accurately predicts the composition of the solvent when density is known. Density and viscosity increases can affect the throughput in the centrifugal contactors. Extremely high density or viscosity can result in a complete loss of solvent flow. The distribution coefficient can also increase, especially in the strip stage, causing the loss of the ability to strip extracted cesium from the solvent. These effects can be addressed by internal changes to the contactor and by adding additional stripping stages in processing. However, these changes are extremely difficult under remote operation and maintenance restrictions (Harmon et al. 2002a).

Caustic-Side Solvent Extraction Engineering Tests of Equipment

Centrifugal contactors have been adopted for the CSSX process due to their small size relative to throughput and short residence time (which reduces solvent degradation) (Fig. 24). As a result of significant testing

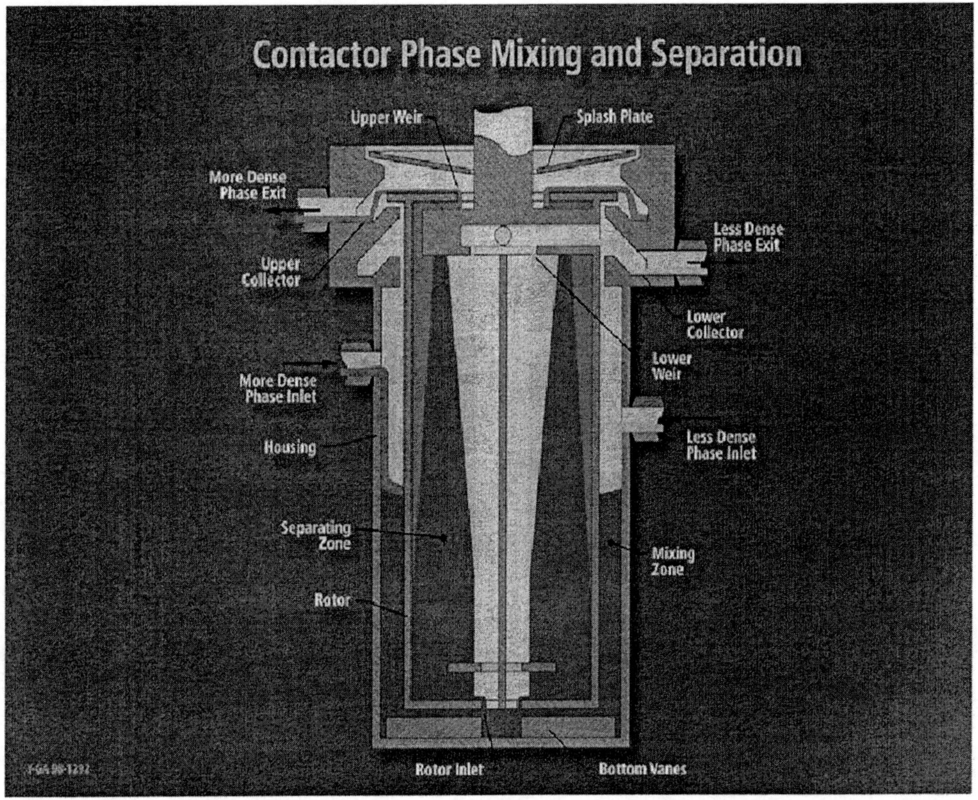

Fig. 24. Centrifugal Contactor Equipment (Leonard et al. 2001a).

of the CSSX process using 2-cm contactors, two engineered equipment issues were identified that required testing with larger-scale equipment to confirm operability, and consequently to reduce technical risk and/or cost of the CSSX process (Harmon et al. 2002b):

1. The first issue involved a scale-up testing evaluation since previous full-system testing had been performed at a small scale (2-cm contactor) and confirmation of larger-scale contactors was required. The risk of larger contactor operation was considered to be low. Modified design criteria for larger contactors were expected to ensure high-stage efficiencies.

2. The second issue focused on solvent recovery due to losses associated with the centrifugal contactor operation. This is a cost minimization issue associated with recovering an expensive solvent. However, it does not impact the operability of the CSSX process. There are two routes for solvent loss in the aqueous phase: 1) natural partitioning (via solubility) of the solvent components between the aqueous and organic phases; and 2) entrainment (or entrapment) as droplets dispersed within the aqueous phase. The amount of solvent lost as entrainment depends on the plant operation (including process upsets that could cause this loss to increase). It was expected that the amount of solvent lost via entrainment would be much larger than that lost via solubility. Therefore, the solvent recovery efforts focused on minimizing entrainment as the primary route for reducing solvent loss. The testing focused on three recovery options: 1) a decanter; 2) a contactor stage using the diluent; and 3) a coalescer.

Leonard et al. (2001a) measured the hydraulic performance of a 5-cm centrifugal contactor from Costner Industries Nevada Corporation (CINC) for both one- and two-phase flow (Fig. 25). The liquid height in the annular mixing zone and the occurrence of discontinuous (slug) flow in the interstage lines were also reported. The flow conditions and test liquids were chosen so as to be appropriate for a CSSX process. In each case, one-phase flow tests were used to predict the behavior in two-phase flow. This way the one-phase flow test can be used as a quality control tool for evaluating contactor rotors as they are manufactured. The results were compared with earlier results obtained using 2- and 4-cm contactors made at Argonne National Laboratory. These results indicated that the 5-cm contactor works in the same way as the 2- and 4-cm contactors.

Birdwell and Anderson (2002) conducted tests to determine if satisfactory mass transfer performance could be achieved using a fully pumping 5-cm centrifugal contactor under conditions present in the CSSX process. The tests utilized a commercially-available contactor that had been modified by installation of a rotor housing bottom with straight radial vanes on the process side. As received from the vendor, the housing bottom was equipped with curved (impeller-type) vanes that were intended to promote phase separation by minimizing mixing of influent solutions. Stage efficiencies exceeding 85% were obtained under conditions present in the extraction section of the CSSX flowsheet.

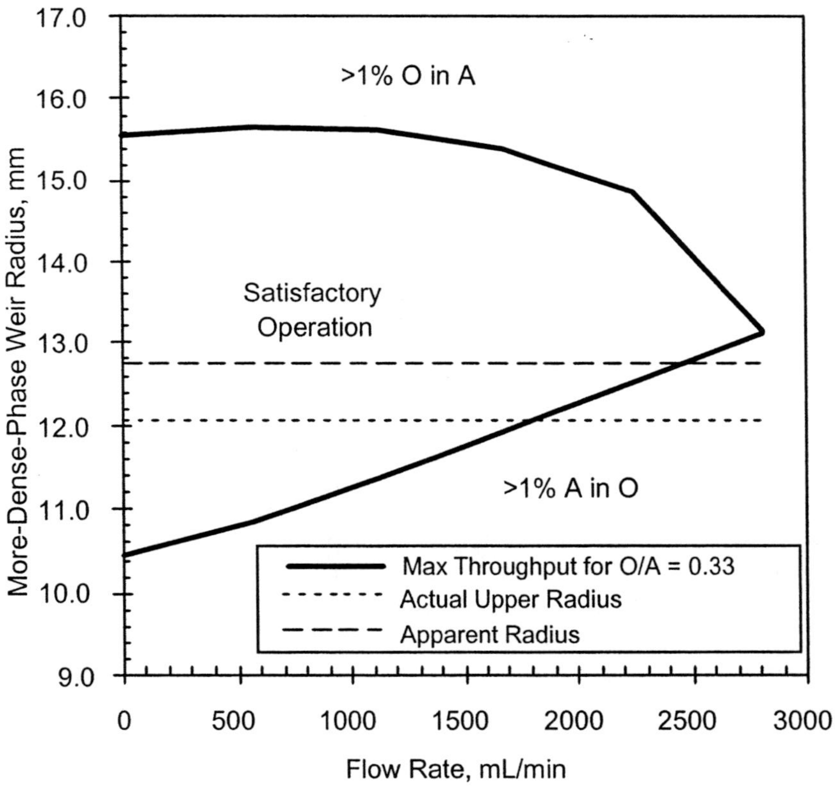

Fig. 25. Maximum Throughput for 5-cm CINC Contactor under CSSX Extraction Section Conditions (Leonard et al. 2001a).

Under CSSX stripping conditions, the stage efficiency exceeded 90%. In both cases, the efficiencies obtained exceed the minimum requirement for acceptable transfer of cesium in the CSSX process.

Law et al. (2002) assessed the hydraulic performance and mass transfer efficiency using a single-stage 5.5-cm ORNL-designed centrifugal contactor. The maximum throughput of the 5.5-cm centrifugal contactor, as a function of contactor rotor speed, was measured for the extraction, scrub, strip, and wash sections of the contactor. These measurements were done using the baseline organic/aqueous flow ratios (O/A), as well as at O/A's 20% higher and 20% lower than the baseline. The maximum throughputs were comparable to the design throughput of the contactor, as well as with throughputs obtained previously in a 5-cm centrifugal contactor using the CSSX solvent prior to optimization. The 20% variation in O/A had minimal effect on the contactor throughput. Additionally, mass transfer efficiencies have been determined for the extraction and strip sections of the flowsheet. The efficiencies were lower than the process goal of 80%, ranging from 72-75% for the extraction section, and from 36-60% in the strip section.

Arafat et al. (2001) evaluated a decanter to recover the solvent from the aqueous effluents. The solvent entrainment was measured in the raffinate and strip aqueous effluents in the CSSX process. Values were obtained for both the solvent concentration and its droplet size distribution. The mixing intensity of the two phases in the mixing zone of the contactor was used to simulate the performance of laboratory-scale, pilot-scale, and plant-scale contactors. The droplet size distributions were used to estimate the amount of solvent that would be recovered using a decanter tank. It was concluded that the performance of decanter tanks would not be as effective in solvent recovery in the CSSX plant as that of other equipment, such as centrifuges and coalescers. Future testing was recommended to verify the performance for the alternative equipment.

Arafat (2002) tested the solvent recovery from the aqueous CSSX process raffinate effluent using the process diluent (Isopar® L). A model was developed to obtain stage efficiency for the diluent contact stages. This model was used to fit experimental data from a 19.8-hour solvent recovery test. Diluent-in-aqueous entrainment was measured during the test. Vacuum distillation was used to concentrate the solvent components

(BOBCalixC6 and modifier) in the diluent used in solvent recovery. Using the results, a feasibility study was performed to compare the annual cost of lost solvent when the four solvent recovery options are adopted: 1) decanter tank; 2) centrifuge; 3) 2-stage diluent contact using centrifugal contactor; and 4) no solvent recovery action.

After CSSX treatment, a small fraction of the solvent is entrained in the caustic raffinate at a level of 100-300 ppm that is well above the solubilities for the various solvent components. Recovery of this solvent could result in significant cost savings (Harmon et al. 2002b). Estimating the physical properties of the solvent and simulant as a function of temperature, Pereira et al. (2002) provide a preliminary evaluation of a coalescer for solvent recovery. Based on laboratory-scale coalescer testing, 80-90% recovery of the entrained solvent can be achieved using a commercial coalescer equipped with appropriate media and at appropriate operating conditions.

On the basis of relevant information, Harmon et al. (2002b) conclude that:

1. neither the decanter nor the contactor using diluent has demonstrated cost-effective solvent recovery.
2. physical property data (Weber Number) indicate that the solvent should be recoverable using a coalescer.
3. laboratory-scale coalescer recovery ranged widely, but 80-90% removal is expected for a properly-designed unit.
4. coalescer media selection is important; tests indicate that stainless steel is a good candidate for the process conditions.

Caustic-Side Solvent Extraction Process Chemistry

Using cesium distribution ratio data from the Oak Ridge National Laboratory (ORNL) on the candidates for the optimized solvent, Leonard (2001) made calculations to determine how each solvent would perform in the CSSX process for both the current solvent flow rate and the optimum solvent flow rate. The D_{Cs} values for the new solvents to replace the current baseline CSSX solvent (0.010 M BOBCalixC6, 0.50 M Cs-7SB, the modifier, and 0.001 M TOA in Isopar® L) were compared using the current baseline CSSX flowsheet; and the same flowsheet with a solvent flow rate optimized for the particular solvent. Except for the two solvents with 0.010 M TOA, all solvents met the robustness criterion at the baseline solvent flow rate. When the optimized flow rate for each solvent is used, all solvents meet the robustness criterion (Klatt et al. 2002). Since the temperature and stage efficiency assumptions are conservative, any of these solvents would work well with respect to flowsheet robustness for the CSSX process. Thus, the final solvent selection was based on other solvent properties including viscosity, dispersion number, third-phase formation, density, and BOBCalixC6 solubility.

The potential for criticality in the proposed SWPF is low but must be addressed in support of Authorization Basis development. Previous tests with simulant showed low extraction of actinides. The results of testing the solvent extraction process with varied solvent compositions show the following (Wilmarth et al. 2002a):

1. During the extraction process, no tested solvent composition showed uranium concentrations in the organic phase above minimum detection limits.
2. Sporadic samples showed plutonium in the organic phase from several of the tested organic compositions indicating probable entrainment rather than extraction.
3. Contact between the solvent and a plutonium-traced acid stream, indicative of the scrub acid, showed little transfer of activity to the organic phase. Plutonium distribution coefficients were less than 0.0006.
4. The data indicate that extraction of actinides into the organic phase will not occur. The actinides prefer the aqueous phase in the scrub stages.

Models show that uranium and plutonium concentrations can exceed the predicted solubilities when scrub acid solution (0.2 M nitric acid) mixes with alkaline waste solutions in the CSSX stages of the planned SWPF. However, in all cases, including the simultaneous precipitation of both uranium and plutonium, the quantities of precipitated uranium and plutonium solids are well-below the fissionable mass limits for ^{235}U and ^{239}Pu (Hobbs 2002). Under normal operations at nominal contact volumes and times, the prediction models indicate that uranium would not precipitate in the CSSX stage, but could precipitate after mixing with large volumes of scrub acid solution and extended storage in the decontaminated waste solution storage tanks. The maximum quantity of uranium precipitated in the full-scale facility was estimated at 396 g (diluted waste volume of 100,000 gallons; 380,000 L). These quantities are well-below the ^{235}U mass limit of 700 g. Under normal operations at nominal contact volumes, plutonium precipitation could occur during the CSSX stage if the feed waste solution enters saturated with plutonium. However, the maximum quantity of plutonium precipitated in the full-scale facility was estimated at 150 g at the fully saturated condition. These quantities are well-below the fissionable mass limit for ^{239}Pu of 450 g. In actuality, the quantities of plutonium will be lower than this value since >90% of the plutonium in the feed waste solution will be removed during the actinide and strontium separation process prior to processing in the CSSX stage.

Leonard et al. (2002) tested the optimized solvent for a CSSX process that removes cesium from SRS tank waste. The optimized solvent was evaluated using the same CSSX flowsheet; SRS simulant; and 33-stage minicontactor (2-cm centrifugal contactor) that had been used to test the previous baseline CSSX solvent. As with the previous solvent, the key process goals were achieved as follows:

1. Cesium was removed from the waste with decontamination factors >40,000.
2. The recovered cesium was concentrated by a factor of 15 in dilute nitric acid.

Thus, the optimized CSSX solvent can be used in place of the baseline solvent while maintaining satisfactory hydraulic performance and still achieving process requirements at the bench scale. The results of this test and previous simulant flowsheet tests are summarized in Table 8.

Table 8. Summary of CSSX Flowsheet Tests Using Simulant (Leonard et al. 2001b).

Test	Solvent Wash	CalixC6 Conc, mM	Cs-7SB Conc, M	TOA Conc, mM	Test Length, h	Solvent Cycles	Cs CF	Cs DF
1	No	10	0.50	1	1.7	Not recycled	16.0	69,000
2	No	10	0.50	1	3.0	4.4	14.6	82,000
3	Yes	10	0.50	1	70.6	42.7	14.9	195,000
4	Yes	7	0.75	3	13.8	11.2	15.5	420,000

Maskarinec et al. (2002) assessed the compositional changes in the CSSX solvent used in the multiday flowsheet test of the CSSX process performed at Argonne National Laboratory (ANL) by Leonard et al. (2001b). The concentrations of the solvent components were determined with two new analytical methods with improved qualitative capability and increased quantitative accuracy and precision. The improved methods can detect small changes in solvent composition or the retention of organic compounds that were in the simulant waste feed. The results are based on comparison of the Argonne multiday solvent sample with a sample of the pristine solvent archived at Oak Ridge National Laboratory (ORNL). The pristine solvent consisted of 0.010 M BOBCalixC6, 0.5 M Cs-7SB, and 0.001 M TOA in Isopar® L. Qualitative and quantitative analyses—carried out to compare the pristine solvent with that used in the Argonne multiday test—showed no unexpected components. The only significant differences between the samples were the increase in dioctylamine concentration at the completion of the test (which was expected), and a slight increase in the modifier and calixarene concentrations. These results indicate that the solvent system did not deteriorate during this limited period of use, and that the small changes in solvent composition had no impact on the results obtained from the ANL multiday test. However, an increase in both calixarene and modifier concentrations suggests that some evaporation of the Isopar® L diluent occurred during the ANL multiday test.

Batch distribution experiments indicated a potential cesium-stripping problem for the CSSX process with high nitrite concentration in the waste simulant (Harmon et al. 2002c). Experiments conducted by Delmau et al. (2002b) have demonstrated that the reason for the cesium-stripping problem was the presence of an anti-caking agent in the sodium nitrite used for the preparation of the simulants. The anti-caking agent is actually a mixture of well-known surfactants (sodium mono- and di-methyl naphthalene sulfonate) that can partition into the organic phase on extraction, then retain cesium upon stripping. The effect was demonstrated by adding known amounts of the anti-caking agent to clean systems. Data show that rejuvenation of the solvent is obtained by a caustic wash following the stripping stages.

Delmau et al. (2002c) extended the model developed for predicting equilibrium distribution ratios in the CSSX process (Fig. 26). This model predicts the extraction performance of the optimized solvent composition, and involves the extraction of cesium and potassium from different cesium, potassium, and sodium media over a large range of concentrations. These different media include a large variety of anions such as nitrate, hydroxide, nitrite, chloride, fluoride, sulfate, and carbonate. The model was derived from several hundred experimental data points, and predicted satisfactorily the cesium extraction from five different SRS waste simulants. Fluoride, sulfate, and carbonate species were found to have low extractability, and their main impact is reflected through their activity effects. The capability of this model to predict cesium and potassium extraction behavior from sodium salts is one of its main features needed for the simulation of actual waste solutions. The extraction from potassium or cesium salts and the extraction of sodium could be improved, and some effort was devoted to improving the thermodynamic rigor of the model. More detailed anion-specific models were developed based on the cesium, potassium, and sodium distribution ratios obtained with simple systems containing single anions, but it has not yet proven possible to combine those models to obtain better predictions than provided by the simpler process model (Harmon et al. 2002c).

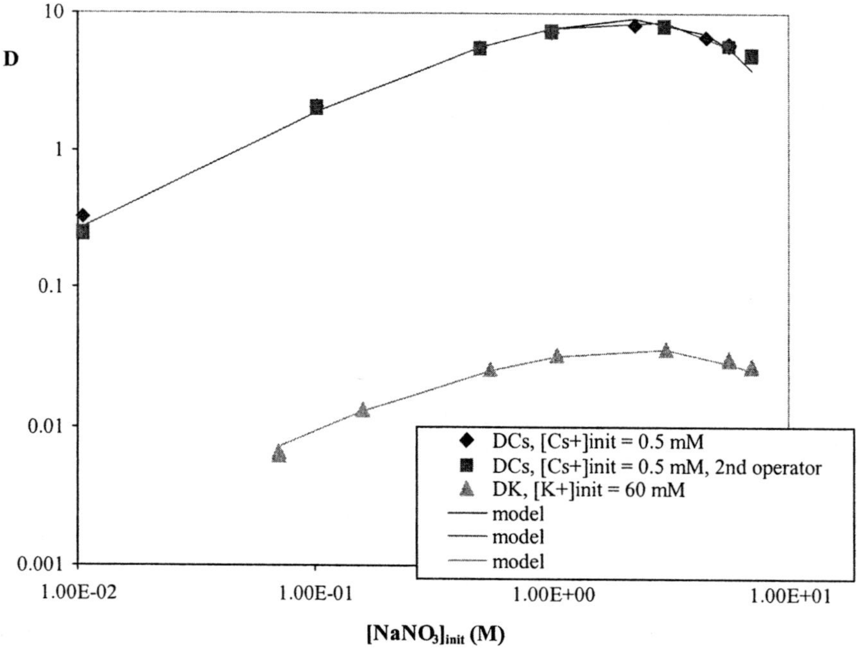

Fig. 26. Modeling of Cesium and Potassium Distribution Ratios for the Nitrate Media (Delmau et al. 2002c).

Birdwell et al. (2002) evaluated the effects on phase separation and emulsion formation of variables present in the caustic washing of solvent in the CSSX process. The evaluation program was performed in two experiments; results of the first experiment were used to determine conditions for the second tests. In the first experiment, the effects of solvent degradation product concentrations (4-sec-butylphenol and dioctylamine); wash solution sodium hydroxide concentration; and solvent-to-wash solution volume ratio (O/A) on phase separation were examined. The phase separation performance was quantified in terms of the dimensionless dispersion number, a variable used in the prediction of centrifugal contactor performance. In the second experiment, the phase separation performance in a 5-cm centrifugal contactor was investigated as a function of contactor speed, aqueous-phase NaOH concentration, and solvent-to-wash volume ratio. The separation performance was quantified in terms of the maximum throughput that could be achieved. Results of the experiments indicated that:

1. At 95% significance level, none of the variables considered affected phase separation performance.
2. At 90% significance level, dioctylamine concentration was the only single factor that affected phase separation.

The results do indicate a 90% significance for interaction effects between NaOH concentration and O/A, and between NaOH concentration and 4-sec-butylphenol concentration. However, none of the dispersion numbers obtained were inconsistent with values obtained in previous studies of CSSX solvent formulations. Contactor performance tests indicated acceptable phase separation at relatively high throughputs over a range of O/A ratios and wash solution NaOH concentrations (Fig. 27). Emulsion formation during contactor operation was observed only under off-normal conditions. The results demonstrate that from the perspective of phase separation efficiency, CSSX solvent washing can be performed over a range of conditions. Consequently, contaminant removal considerations should not be affected by hydraulic performance concerns in the selection of parameters for solvent treatment (Harmon et al. 2002c).

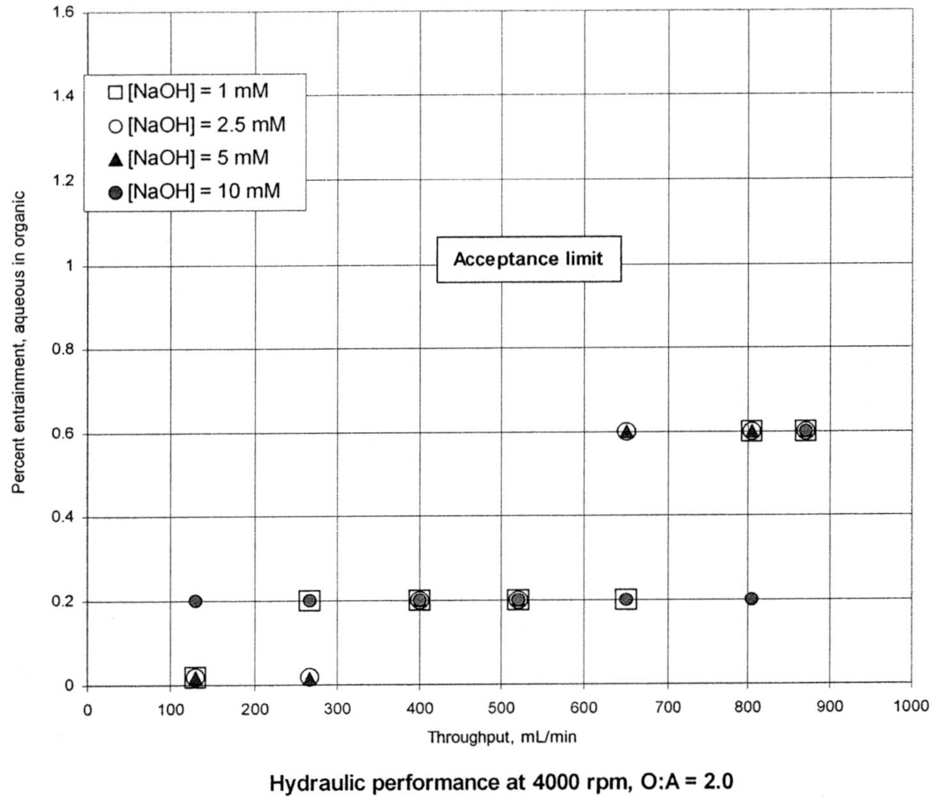

Fig. 27. Centrifgual Contactor Hydraulic Performance During Solvent Washing (4000 rpm, O:A = 2.0) (Birdwell and Anderson 2002).

Bonnesen et al. (2002) investigated whether nitrated organic compounds could be formed during operation of the CSSX process, and whether such compounds could present a safety concern. During the CSSX process, the solvent will be exposed to high concentrations of nitrate and nitrite in the alkaline waste feed. The solvent will also be exposed to dilute (50 mM) nitric acid solutions containing low concentrations of nitrite during scrubbing, followed by stripping with 1 mM nitric acid. The solvent is expected to last for at least one year of plant operation, and the temperatures that the solvent may experience during the process could range from as low as 15°C to as high as 35°C. Excursions from standard process conditions could result in the solvent experiencing higher temperatures, as well as concentrations of nitrate, nitrite, and most importantly, nitric acid that exceed the normal operating conditions. Accordingly, conditions may exist where nitration reactions involving the solvent components, possibly leading to other chemical reactions stemming from nitration reactions, could occur. The solvent was exposed to nitrate- and nitrite-containing solutions that are expected to be encountered during the process (even under off-normal conditions). Based on the results of Bonnesen et al. (2002) and Moyer et al. (2002), Harmon et al. (2002c) conclude that the formation of nitrated organic compounds (under the conditions that solvent is expected to experience during normal plant operations) is essentially non-existent.

Stock et al. (2002) reviewed the technical reports dealing specifically with the organic constituents in the CSSX process for insight on the reactivity under operating conditions. However, most information comes from the literature on synthetic methods development; mechanism elucidation; structure-reactivity studies; and chemical inferences about known reactions of related compounds. This approach was required because little published information is available about the chemistry of the calixarene and the modifier. Although uncertainty remains, the chemical literature provides sufficient information about the chemical reactions of related compounds to establish the probable reactivity patterns of the modifier, calixarene, and TOA. This literature survey has led to the identification of many slow radiolytic and chemical reactions that will occur during the operation of the CSSX process. The reactions of the modifier, calixarene, and TOA and the radiolysis of the hydrocarbon solvent will form many different organic compounds present in low concentrations. Where quantitative rate comparisons are possible, the information in the literature indicates that no single chemical reaction is rapid enough to consume 1% of the modifier, calixarene, or TOA. Unfortunately, rate data are sparse and it is unclear whether or not the array of reactions that have been identified will consume 1% of a CSSX solvent constituent in one year. However, it is certain that many new organic compounds will be formed in low-yield and will be incorporated into the solvent during continuous operation.

Accurate characterization of individual waste components is critical to ensure design and operation of effective treatment processes and compliance with waste acceptance criteria. Current elemental analysis of organic matrices consists of conversion of the organic sample to aqueous by digesting the sample. In many cases this approach is inadequate. Direct analysis of the organic compounds would increase sensitivity and decrease contamination and analysis time. Pennebaker et al. (2002) evaluated an Aridus membrane-desolvation sample introduction system for the direct analysis of organic solvents by Inductively Coupled Plasma-Mass Spectrometry (ICP-MS). The desolvator-ICP-MS successfully analyzed solvent from the CSSX process and tri-butyl phosphate organic tank waste for a variety of elements. Detection limits for most elements were determined in the part per trillion range, which compares favorably with previous detection limits of ICP-MS. This technology could increase accuracy in support of Savannah River Site activities involving CSSX and other site processes involving organic compounds.

Caustic-Side Solvent Extraction Actual Waste Tests

Walker et al. (2002) characterized HLW derived from dissolved salt cake from the SRS tank farm and measured the cesium distribution coefficients (D_{Cs}) for ESS steps of the CSSX flowsheet. The measurements used the old solvent composition and two SRS HLW samples derived entirely or in part from saltcake. Dissolved saltcake waste contained less ^{137}Cs and more di-anions than is typical of supernate samples. Extraction, scrub, and strip D_{Cs} values for both samples exceeded the process requirements. The model predictions agreed well with the measured D_{Cs} values. The results indicate no significant problems regarding the processing dissolved saltcake as compared to supernate. During the course of testing, Walker et al. (2002) observed sorption of cesium on glass sample vials containing strip solutions. The problem was detected in the material balance calculated for each organic/aqueous contact. Methods and recommendations for avoiding this problem were assessed.

The first measurements of cesium distribution behavior in actual waste samples during the CSSX process were reported using the old solvent composition. These measurements showed acceptable behavior across the ESS stages of the CSSX process. However, extraction performance was not consistent with developed thermodynamic models (Harmon et al. 2002d). Wilmarth et al. (2002b) performed additional batch tests using the optimized solvent to measure the distribution coefficients with samples from SRS F- and H-Area HLW tanks. The ESS behavior with actual tank waste (including Tank 37H dissolved saltcake) and the newly-optimized solvent system surpassed flowsheet requirements (Fig. 28). The thermodynamic model predictions of extraction distribution coefficients agreed well with measured values in the optimized solvent system. The ESS distribution coefficients for a waste sample that had been processed using a permanganate treatment to remove transuranic components were acceptable compared to flowsheet requirements. The experimental techniques and counting methodologies greatly improved the ability to perform ESS testing with actual tank waste samples in the Shielded Cells.

Norato et al. (2002a) demonstrated the chemistry of the CSSX flowsheet for the decontamination of HLW using optimized solvent in a 33-stage, 2-cm centrifugal contactor apparatus (Fig. 29). This represents the second CSSX process demonstration using SRS HLW (Harmon et al. 2002d). Three tests were performed using radioactively-"spiked" simulated Tank 37H/44F composite waste, and a 24-hour demonstration was performed using actual Tank 37H/44F composite HLW supernate. The CSSX process using optimized solvent reduces ^{137}Cs in HLW supernate solutions to concentrations below the Saltstone waste acceptance criterion of 45 nCi/g. Waste decontamination factors (DF) as high as 1.37 million resulted during testing. The average DF for the composite waste equaled 520,000 over 24 hours of operation (Fig. 30). Testing occurred over 24 hours of uninterrupted operation, demonstrating hydraulic stability of the contactor array while operating with the optimized solvent formulation. Tests with actual Tank 37H/44F composite waste demonstrated extraction and strip section stage efficiencies at 80%. Carryover of organic solvent in aqueous streams (and aqueous in organic streams) proved less than 1% when processing Tank 37H/44F HLW. Minor components in the high-level radioactive waste did not affect the ability of the contactors to efficiently separate the solvent and aqueous phases.

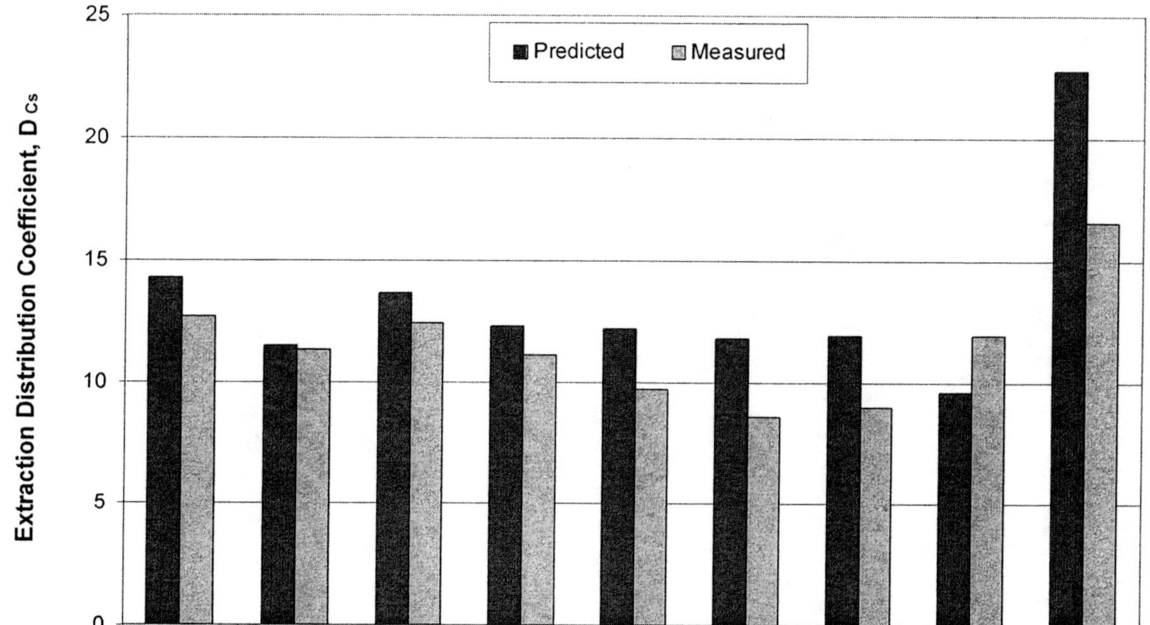

	26F	34F	46F	30H	32H	41H MnO2 Treated	37H SC	38H SC	46F SC
Predicted	14.3	11.5	13.7	12.3	12.2	11.8	11.9	9.6	22.8
Measured	12.7	11.3	12.4	11.1	9.7	8.6	9	11.9	16.6

Fig. 28. Cesium Distribution Coefficients Measured with Actual Waste Samples (Wilmarth et al. 2002b).

Fig. 29. Centrifugal Contactors Used in Savannah River Technology Center Actual Waste Tests (Norato et al. 2002a).

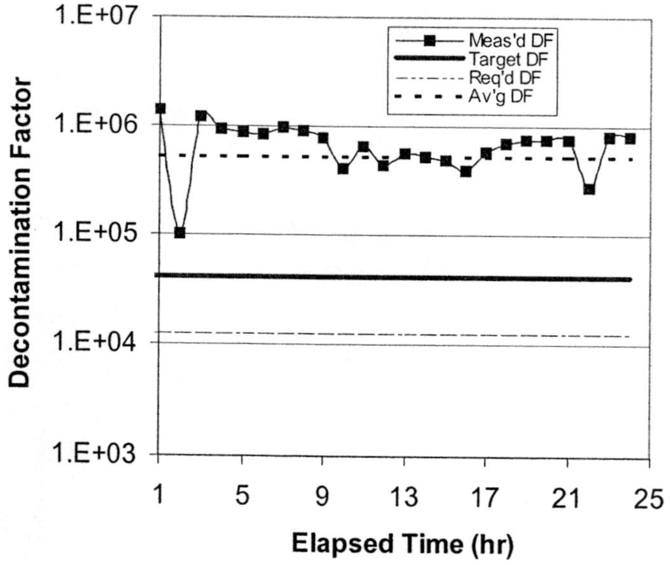

Fig. 30. Cesium Decontamination Factors Measured during Tank 37H/44F Actual Waste Demonstration (Norato et al. 2002a).

The concentration factor (CF) averaged 12.7 over 24 hours of operation with Tank 37H/44F composite solution. Uncertainties in process flow rate measurement and control prevented the test from achieving the target CF of 15.

Norato et al. (2002b) demonstrated the chemistry of the CSSX flowsheet with optimized solvent and dissolved salt solution. This represents the third CSSX process demonstration using actual SRS high-level waste (Harmon et al. 2002d). The present test differed from the previous tests of Campbell et al. (2001) and (Norato et al. 2002a) in the use of radioactive waste derived from Tank 37H dissolved saltcake, as opposed to supernate solutions as used in the two previous demonstrations. The dissolved saltcake waste was taken early in the salt dissolution cycle, and it was found to be similar in composition to the supernate waste previously tested. The test used a 33-stage, 2-cm centrifugal contactor apparatus in a shielded facility at the Savannah River Technology Center (SRTC). The test with actual Tank 37H waste processed 44.6 L of solution in 25.5 hours. This test was preceded by a 4-hour test with simulated Tank 37H waste. The CSSX process using optimized solvent reduces ^{137}Cs in dissolved saltcake solutions to concentrations below the Saltstone process requirement of 45 nCi/g. The results of Waste DF as high as 3 million during testing were: 1) the DF for the composite waste equaled 352,000 over the first 16 hours and 28,600 for the remaining 9.5 hours of testing; and 2) an intentional decrease in the solvent flow rate caused the change in DF after 16 hours. The solvent was stripped of cesium and recycled with an average solvent DF of 22,500 during the first 16 hours and 20,700 during the remaining 9.5 hours of the test. Testing occurred over 25.5 hours of uninterrupted operation, demonstrating the hydraulic stability of the contactor array over a range of process conditions. Tests with actual Tank 37H waste demonstrated extraction and strip section stage efficiencies of 90% and 60%. Carryover of organic solvent in aqueous streams (and aqueous in organic streams) proved less than 1% when processing Tank 37H HLW. Minor components in the high level radioactive waste did not affect the ability of the contactors to efficiently separate the solvent and aqueous phases. The CF averaged 13.2 during the first 16 hours of operation with Tank 37H-dissolved saltcake solution and 15.8 in the remaining 9.5 hours (Fig. 31). Uncertainties in process flow rate measurement and control prevented the test from achieving the target CF of 15 during the initial portion of the test. The results from all three CSSX actual waste demonstrations are summarized in Table 9.

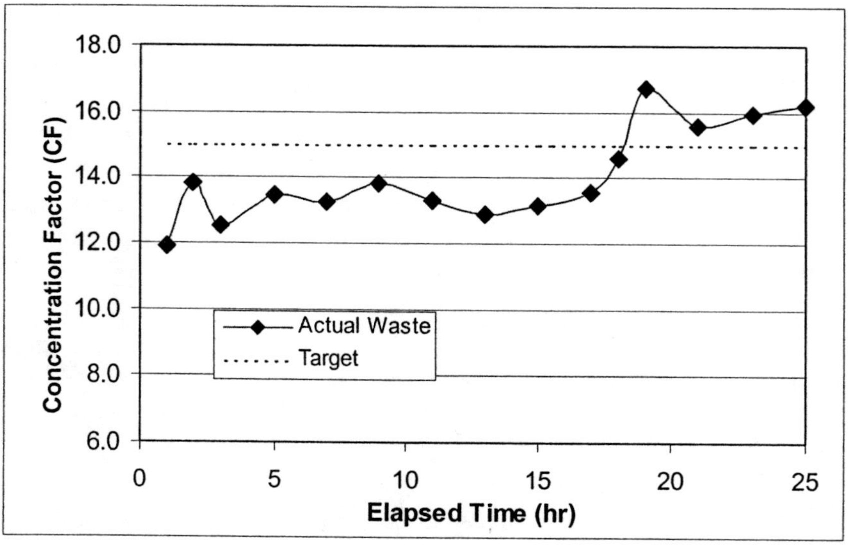

Fig. 31. Concentration Factors Measured During Tank 37H Dissolved Salt Actual Waste Demonstration (Norato et al. 2002b).

Table 9. Summary of CSSX Flowsheet Tests Using Actual Waste (Norato et al. 2002a).

Parameter	Goal	37H/44F Orig. Solv. Supernate	37H/44F Opt. Solv. Supernate	37H Opt. Solv. Dissolved Salt
DF	40,000	511,000	520,000	352,000
CF	15	12.7	12.7	13.2
% Eff. Extr	80	84	90	95
% Eff. Strip	80	82	80	85
Solv. Carryover	<1%	<0.4%	<0.2%	<0.2%
Test Cycle	24 hrs.	34 hrs.	24 hrs.	25 hrs.

Walker (2002a) investigated the stability of SRS radioactive waste solutions in two laboratory experiments. The first experiment tested four waste solutions for supersaturation of aluminum by monitoring the aluminum concentration after seeding with gibbsite. The second experiment tested two waste samples for precipitation of aluminosilicates by heating the solutions to accelerate solids formation. Supersaturation with respect to aluminum occurred in some SRS radioactive waste solutions. The aluminum concentration in Tank 32H radioactive waste solution decreased by 37% over 115 days when seeded with gibbsite; the soluble aluminum concentration in Tank 30H waste did not change; and in Tank 34F and Tank 37H/44F composite waste, the soluble aluminum concentration increased when seeded with gibbsite. Heating did not precipitate detectable amounts of aluminosilicates in waste samples from Tank 38H and 43H. No solids were detected either by visual inspection or by analysis of acid solutions used to clean the apparatus after heating. The evaluation of the Tank 38H and 43H compositions by a Geochemist Workbench solubility model suggests that solids should not have formed during the test. Previous studies showed that simulated waste solutions exhibit instabilities (Walker 2000). The absence of aluminosilicate formation in actual waste samples should not be taken as an indication that this cannot occur during waste processing. Aluminosilicate formation in SRS evaporators is well documented (Wilmarth et al. 1997; Wilmarth et al. 2000a&b). The current flowsheet specifications for waste dilution are inadequate to prevent gibbsite formation in most tank wastes and greatly underestimate the amount of dilution required. Since aluminosilicates and gibbsite form slowly and therefore may not accumulate rapidly in the contactors, nitric acid cleaning of contactors appears to be a less-expensive alternative to waste dilution (Harmon et al. 2002d).

Certain categories of organic compounds can interfere with the CSSX process. Knowledge of the presence and concentrations of interfering compounds in SRS soluble waste is needed to ensure good process performance. The SRS HLW tank farms store and process high-level liquid waste from a number of site sources. The waste has been extensively analyzed for inorganic components but few analyses of organic compounds in waste have been obtained. Based on a review of previous sample analyses and records of disposals of organic compounds of the HLW tank farms, Walker (2002b) identified organic compounds or classes of organic compounds that may occur in SRS HLW.

These classes include tributylphosphate (extractant used in F- and H-Canyons); siloxanes and siloxanols (breakdown products from defoaming agents); ion exchange resin fragments; and alkyl-aryl sulfonates (found in soaps, detergents, surfactants, and certain process chemicals). With the present available information, it is not possible to determine the concentration (and, therefore, the potential threat to the CSSX process) of the interfering compounds. Campbell et al. (2002a&b) developed analytical methods suitable for HLW for the classes of compounds listed above.

Peer Review Criteria, Findings, and Recommendations of the Review Panel

The findings of the RP with respect to the review criteria are as follows:

Criterion 1

Has the Caustic Side Solvent Extraction (CSSX) technology for cesium removal reached the necessary technical maturity required for preliminary design for deployment at Savannah River Site (SRS)?

Finding 1 of the RP

The CSSX technology for cesium removal has reached the necessary technical maturity required for preliminary design for deployment at SRS.

The CSSX was selected as the salt processing alternative primarily because it presents a low technical risk for successfully removing cesium from radioactive waste from salt solutions. The proposed CSSX uses a complex four-component solvent system. Laboratory testing with centrifugal contactors, and a limited number of tests with a highly radioactive waste, have shown that component concentrations and process-flow can be maintained to effectively remove cesium from the wastes. Other key strengths identified for the CSSX technology include:

1. Maturity of and experience with solvent extraction within the DOE Complex. Solvent extraction is widely used commercially in high-radiation environments throughout the world.
2. Simplicity with which the CSSX product stream could be incorporated into the current Defense Waste Processing Facility (DWPF) vitrification process. The acidic cesium waste stream is compatible with DWPF operation.
3. The ability to rapidly start up and shut down the centrifugal contactors, coupled with the short residence time of the solvent in the contactors, which minimizes the solvent degradation.

Criterion 2

Has the monosodium titanate (MST)/Filtration technology for removal of actinides and strontium reached the necessary technical maturity required for preliminary design for deployment at SRS?

Finding 2 of the RP

The MST/Filtration technology for removal of actinides and strontium has reached the technical maturity required for preliminary design for deployment at SRS. Sorption and filtration technologies are well established throughout the industrial world.

MST removal of actinides and strontium results in compliance with the Saltstone Waste Acceptance Criteria for disposal as low-level waste. The experimental Decontamination Factors (DFs) for strontium exceeded the required minimum DF for saltstone disposal.

The experimental DFs obtained for plutonium indicate that satisfactory removal can be achieved, provided that sufficient MST is used in the feed. Performance can be further enhanced through multiple strikes and interim filtrations. The concentrations of uranium, neptunium, americium, and curium in the waste salt are sufficiently low to make further removal unnecessary for saltstone disposal.

Recent studies demonstrate that with respect to plutonium removal, MST performs better than the principal competing inorganic sorbents that are either currently available at commercial scale or in final stages of development.

The Project Team also assessed the merits of other alternatives (including sodium permanganate) to MST to remove soluble strontium and actinides from SRS high-level waste. The assessment for sodium permanganate considered deployment in both the Salt Waste Processing Facility (SWPF) and within Building 512-S for the Actinide Removal Process (Fink et al. 2002a). The alternative technologies investigated by the Project Team lack sufficient maturity for deployment.

Criterion 3

Has the Project Team identified the residual technical risks associated with implementing the CSSX technology? In particular, has the Project Team adequately defined the operating conditions and limitations of the different unit operations involved in this technology?

Finding 3 of the RP

The Project Team has identified the residual technical risks associated with implementing the CSSX technology and concluded that the identified risks are sufficiently low to proceed with preliminary design of the SWPF. The Project Team has adequately defined the operating conditions and limitations of the different unit operations involved in this technology.

Chemical and physical properties of the solvent

A novel feature of the CSSX system is the use of a multicomponent solvent consisting of:

1. 0.007 M calix[4]arene-bis(tert-octylbenzo-crown-6), also called BOBCalixC6, which is the extractant
2. 0.75 M 1-(2,2,3,3, -tetrafluoropropoxy)-3-(4-sec-butylphenoxy) -2-propanol, also called Cs-7SB, which is the modifier
3. 0.003 M tri-*n*-octylamine (TOA), which is the suppressor
4. Isopar® L, which is the diluent

This "optimized" solvent composition, arrived at after extended testing at the Oak Ridge National Laboratory (ORNL), has a high specificity for potassium and cesium ($D_{Cs} > 9$ at 25°C), and the cesium can be stripped from the solvent with dilute HNO_3. The extractant was originally synthesized at ORNL, and kilogram quantities can now be obtained from at least one commercial supplier. The following properties, important in the design of the solvent extraction systems, have been determined (Klatt et al. 2002):

1. Distribution coefficients for cesium
2. Solubilities in the diluent and in the aqueous solutions
3. Third phase formation (not present)
4. Dispersion characteristics
5. Density
6. Viscosity
7. Surface and interfacial tensions
8. Radiation, chemical, and thermal stabilities
9. Volatility

In addition, several tests were run to determine the effects of impurities and solid particles on some of the above listed physicochemical properties. The results showed no significant effects within the ranges investigated.

Radiation stability is of paramount importance for the processing of highly radioactive materials. Tests using internal and external radiation sources showed that the solvent possesses adequate radiolytic stability for its intended use.

Traces of solvent degradation products can be removed during solvent washing to allow for continuous recycle without affecting the extraction or stripping characteristics. However, it is prudent to consider additional testing using real waste for extended periods of time, preferentially coupled with contactor operation. Periodic replenishment of the tri-*n*-octylamine (TOA) component may be required.

All tests performed so far lead to the conclusion that the proposed solvent will be adequate for the CSSX process.

Process equipment

Centrifugal contactor technology is mature. Application-specific testing of contactors was performed to confirm operability and to reduce the residual technical risks of the CSSX process. Limited tests using 2-cm contactors were performed at the Argonne National Laboratory (ANL). Subsequently, tests were performed using 5-cm and 5.5-cm contactors at ORNL and the Idaho National Engineering and Environmental Laboratory. These tests, which used simulated process solutions, included hydraulic tests under CSSX extraction, scrubbing, and stripping conditions. They also included mass-transfer tests under extraction and stripping conditions. Excellent phase separation was achieved under all flowsheet conditions. The Project Team plans to perform additional large-scale tests during preliminary design. Entrainment in the aqueous phase and evaporation in the head space cause small losses of solvent. The Project Team reported that efficient solvent recovery and recycle are required because the solvent is very expensive. Testing was performed on different methods of recovering entrained solvent. These methods included decanting, centrifuging, coalescing, and diluent scrubbing. During preliminary design, both decanting and coalescing alternatives will be considered.

Process chemistry and criticality concerns

Fundamental to the CSSX process is the extraction of radioactive cesium from a caustic aqueous phase into the multicomponent solvent. In the aqueous phase, this separation is required to reduce the ^{137}Cs concentration by a factor of about 40,000 to allow the incorporation of the decontaminated salt solution in saltstone.

Many experiments at the laboratory and engineering levels have been performed to optimize the concentrations of the solvent components. Hundreds of sets of data have been analyzed and included in the development of a model for the CSSX process chemistry to extract and strip ^{137}Cs from the salt waste feeds. The model accurately predicted the cesium extraction for processing five different SRS salt waste simulants, and it provides an excellent method for predicting cesium and potassium removal from the waste salts (Delmau et al. 2002c).

Criticality concerns about uranium and plutonium have been addressed. The Project Team concluded that uranium and plutonium extraction and accumulation into the organic phase are not significant.

Early observations of inadequate cesium stripping were attributed to impurities present in the waste simulants. An increase in the solvent component tri-*n*-octylamine (TOA) from 0.001 M to 0.003 M mitigates stripping problems.

Numerous tests involving exposure to radiation, temperature variations, and chemical changes proved that the stability of the solvent is very high. The Project Team plans to remove the trace-level degradation products from the solvent by caustic washing. This minimal-effort control effectively eliminates extraction and hydraulics problems that might result from a buildup of degradation products.

The Project Team has identified and addressed the process chemistry risks, and has correctly assigned a low level to these risks.

Testing

Simulant waste: After the solvent composition was "optimized," flowsheet testing was performed with simulant waste feed. The key goals achieved were:

1. Cesium was removed from the simulant waste with DF values greater than 40,000
2. Cesium stripping with dilute nitric acid resulted in a volume ratio of aqueous feed to an aqueous product of 15

Thus, the "optimized" solvent met (at bench and engineering scales) the process design requirements while maintaining satisfactory hydraulic performance. Organic-phase sample analyses indicated that the solvent system did not deteriorate during this limited period of use.

Actual Waste: Batch distribution measurements were conducted with waste samples from SRS F- and H-area HLW tanks. Batch-distribution coefficients indicate that the solvent is suitable for varying waste compositions. Good agreement was observed between measured and predicted values.

Flowsheet tests were conducted in 2-cm centrifugal contactors using both supernate and dissolved salt cake from Tank 37H. These tests demonstrated that the base-line, as well as the "optimized" solvent, reduced ^{137}Cs concentrations to those below the Saltstone Facility process requirement of 45 nCi/g. Waste decontamination factors (DFs) were at least eight times larger than the stated goal of 40,000. The solvent was efficiently stripped of cesium and recycled. The carryover of organic solvent in aqueous streams (and aqueous in organic streams) was less than 1%. Stability tests performed with the SRS radioactive waste solutions indicated that supersaturation, in aluminum, occurs in some solutions—which has resulted in aluminosilicate precipitation in a tank farm evaporator. The kinetics of this precipitation were determined to be slow enough so that filtration, just prior to CSSX processing, is sufficient to avoid any precipitation in both the simulant and actual waste CSSX system tests.

Criterion 4

Has the Project Team identified the residual technical risks associated with implementing the MST/Filtration technology? In particular, has the Project Team adequately assessed the compatibility of the MST/Filtration product stream with the CSSX process?

Finding 4 of the RP

The Project Team has identified the residual technical risks (plutonium sorption and removal) in MST/Filtration technology. The Project Team also adequately assessed the compatibility of the MST/Filtration product stream with the CSSX process.

Pu Sorption and Removal Performance: The Project Team examined the sorption of plutonium and other radionuclides by MST under prototypical conditions. These studies included:

1. Proof of compatibility of plutonium and strontium removal with MST for each one of the three cesium removal process options (STTP, CST, CSSX) using both simulated and actual wastes
2. Tests with simulated waste containing added actinides and strontium
3. Experiments with actual HLW

The accumulated data demonstrated successful operation for a variety of waste compositions while meeting the process requirements defined for the proposed facility. Although the rate of plutonium sorption limits the total processing time, MST adequately removes plutonium with acceptable efficiency.

The Project Team has demonstrated that there is a low probability for failure of this sorption process. Furthermore, the Project Team indicated that remedial actions for any inadequate sorption will consist simply of adding fresh MST. This will result in an insignificant increase in operation time.

MST/Filtration: Cross-flow filtration is a widely-used and well-developed technology. Its successful use in conjunction with MST sorption has been demonstrated with actual waste in full-scale equipment in the SRS In-Tank Precipitation Facility (Harmon et al. 2001). Previous demonstrations included chemical cleaning and backpulsing, the two process steps necessary to ensure prolonged operation. Therefore, there is a very low risk in implementing this technology for the SWPF.

Compatibility: The Project Team has demonstrated the MST/Filtration process product stream to be of sufficiently-low plutonium and solids content to ensure its compatibility with the feed requirements of CSSX.

Criterion 5

Has the Project Team adequately assessed the compatibility of the SWPF product streams (MST/Sludge; CSSX cesium strip effluent; and decontaminated salt solution) with the Defense Waste Processing Facility (DWPF) and Saltstone Facility?

Finding 5 of the RP

The Project Team has adequately assessed the compatibility of the SWPF product streams (MST/Sludge; CSSX cesium strip effluent; and decontaminated salt solution) with the Defense Waste Processing Facility and Saltstone Facility. Specifically, the MST/Sludge can be fed to the DWPF without further treatment, and the properties of the resulting waste glass will not be affected significantly. The CSSX cesium strip effluent is sufficiently concentrated to be fed to the DWPF without additional treatment. The decontaminated salt solution can be transferred to the Saltstone Facility without further treatment.

MST/Sludge: Slurry rheology measurements showed that the relevant parameters are comparable to those encountered in past transfers: suspension and pumping studies showed adequate mixing at operating temperatures. The Project Team has demonstrated that waste glass, containing the equivalent of up to 2 wt% TiO_2, is acceptable. The introduction of MST/Sludge at the projected feed rate will add less than 2 wt% TiO_2 equivalent.

CSSX Cesium Strip Effluent: This acidic waste stream can be fed to the DWPF feed preparation tank without any further concentration. The CSSX process not only removes the cesium from the waste stream, but also increases the concentration of cesium in the product stream by a factor of 15.

Decontaminated Salt Solution: The decontaminated salt solution, after incorporation in grout, is disposed of in the onsite Saltstone Facility. The Project Team is aware of the need to carefully control the cesium content in order to meet existing regulatory specifications. The laboratory and pilot scale tests with actual waste have shown that the SWPF product can meet these specifications.

In summary, in each case it is anticipated that the products from the SWPF will be compatible with the Defense Waste Processing Facility (DWPF) and Saltstone Facility.

Criterion 6

Has the Project Team adequately planned technology development activities that are anticipated to further mature or optimize the CSSX technology during the design or construction stages?

Finding 6 of the RP

The Project Team has adequately planned technology development activities that are anticipated to further mature or optimize the CSSX technology during the design or construction stages. The plans submitted by the two EPCs are not identical, since some activities shown in the plans of one of the EPCs are not shown in the other. The proposed plans show a thorough endeavor to further improve the CSSX process during Phase II, consistent with the developments during Phase I.

Criterion 7

Has the Project Team adequately planned technology development activities that are anticipated to further mature or optimize the MST/Filtration technology during the design or construction stages?

Finding 7 of the RP

The Project Team has adequately planned technology development activities, considering that the MST/Filtration technology is already mature and has been used successfully. The plans submitted by the two EPCs are not identical, since some activities shown in the plans of one of the EPCs are not shown in the other. The proposed plans show a thorough endeavor to further advance the MST/Filtration implementation during Phase II, and are consistent with their developments during Phase I.

The attrition of the loaded MST during the slurry transfer to the DWPF may lead to the generation of plutonium-rich fine particles that could deposit in the transfer pipe. This could create an accumulation of plutonium-rich material, a potential safety issue.

Criterion 8

Has the Project Team adequately addressed the recommendations provided by independent peer reviews and technical assessments?

Finding 8 of the RP

The Project Team has adequately addressed the recommendations provided by the two independent NRC peer reviews (NRC 2000; 2001b).

Caustic side solvent extraction

The NRC (2000) found that although solvent extraction in general is a well-developed process, the technical maturity of the proposed solvent extraction process for the removal of cesium from high-level waste (HLW) at SRS lagged significantly behind its competitors. The NRC (2000) questioned whether the solvent could be cleaned and reused successfully. The buildup of particulate matter at the interface of the two liquid phases, which is undesirable in solvent extraction, was considered to be a possibility in the process at SRS, and therefore it could limit separation of phases. The impact of changes in feed composition, pH, temperature, and ratios of the solvent constituents on the separation efficiency and capacity of the solvent system was also of concern. Other issues included the need for sophisticated control systems required to maintain a steady-state; high-DF operation; and the possible interference of certain ions present in tank waste with efficient separation. The NRC (2000) also questioned whether a reliable supply (and supplier) of the calixarene crown ether (BobCalixC6) will be available. As a result of these concerns, the NRC (2000) recommended:

1. A "cold" demonstration of this process on a modest scale. The purpose of this demonstration is to address as many of the aforementioned operational concerns as possible and, in particular, to identify any possible "showstoppers" that would preclude the use of this process at SRS.
2. A "hot" laboratory demonstration (using real tank waste) on a scale sufficient to define the final process. This demonstration should begin as soon as the cold tests provide sufficient confidence in the feasibility of the process.
3. The definition of production capability and economics for commercial quantities of BOBCalixC6.

The Project Team implemented a robust R&D program that addressed many of the concerns expressed by the NRC (2000). A demonstration of the complete CSSX process flowsheet (using simulated waste) was conducted, and some real waste tests were also performed. The R&D program addressed factors such as: 1) the chemical and physical properties of the solvent; 2) the stability of the solvent system; 3) batch DF extraction and stripping; and 4) procurement of solvent components. Following the above efforts, another review was conducted by the NRC (NRC 2001a&b). The NRC (2001b) panel stated that "there do not appear to be any technical obstacles to scale-up of the CSSX process to plant-scale operations." The successful bench-scale demonstration of the complete process with actual tank waste has been flagged as critical for qualifying the CSSX process for serious consideration in the down-selection process (NRC 2001b). The NRC (2001b) provided the following two recommendations:

1. Monitoring of the cost and potential suppliers of the reagents for this process should continue.
2. Extensive testing of the most performance-critical components of the solvent (e.g., composition, pH, and temperature ranges the solvent would most likely encounter) should continue in parallel with the bench-scale process test using real plant waste.

The Project Team identified one commercial supplier who produced and delivered kilogram quantities of BOBCalixC6. Two commercial suppliers were identified for Cs-7SB. Other suppliers have indicated an interest in contracting for the production of kilogram quantities of these reagents. During the oral presentations, the Project Team indicated that:

1. The syntheses of BOBCalixC6 and Cs-7SB are relatively straight forward
2. Ownership of the basic patents by DOE provides protection against excessive future costs
3. The other two solvent components, TOA and Isopar® L, are commercially available

The Project Team performed testing of the most performance-critical parameters of the multicomponent solvent. Three bench-scale process tests using approximately 200 L of real plant waste (supernate or dissolved saltcake) were carried out at SRS. DFs; concentration factors; and extraction and stripping efficiencies were obtained for these integrated experiments.

Monosodium titanate

The NRC (2000) found that two major issues had to be resolved before SRS could successfully implement MST for actinide and strontium removal:

1. Whether strontium and actinide removal could be accomplished within saltstone limits and throughput rates required by the DWPF
2. Whether the MST concentrations used to remove strontium and actinides would exceed compatibility limits for DWPF glass

The NRC (2000) recommended that:

1. R&D be performed to resolve these issues
2. Requirements and reliable sources for the manufacture of MST be established
3. SRS look at alternatives to MST

As a result of the R&D activities performed by the Project Team, the NRC (2001b) found that the blending of tank waste to produce 67 process batches and treatment by MST appears to meet the saltstone requirements for neptunium and strontium decontamination. MST appeared "to be adequate to separate Pu, as long as there is no colloidal plutonium in the waste, but with little margin to meet saltstone requirements" (NRC 2001b). However, the maximum quantity of titanate allowable in the process stream to meet the titanium levels acceptable in the vitrified waste form remained a technical uncertainty. The NRC (2001b) recommended that the "backup processes, sodium nonatitanate and the sodium permanganate-based precipitation process, should be studied further. The R&D program for these two processes should be based on that developed for MST and should continue until MST processing can be demonstrated to meet the saltstone, DWPF throughput, and DWP glass requirements."

The Project Team has demonstrated that the MST/Filtration process meets the saltstone, DWPF throughput, and DWPF glass quality requirements, and no problems have been identified in engineering this process. However, work has continued to identify alternatives that could serve as a backup process should there be any future problems with MST. In the course of these tests, the Project Team found that sodium permanganate-based precipitation lacks sufficient maturity to be deployed. Other sorbents (e.g., sodium nonatitanate; SrTreat®; CST; pharmacosiderites) were tested and showed some promise, but have not reached sufficient maturity.

Additional Finding of the RP: Finding 9

The RP was impressed by the thorough planning; quality of experimental work; risk assessments; and engineering analyses performed by the Project Team with the participation and cooperation of several organizations; and on a tight schedule. It was evident that the Project Team has maintained a sharp focus on the principal objectives.

DOE management is commended for having expanded the Project Team in Phase I through the inclusion of two competing EPC vendors. At the end of Phase I there will be a down-selection so that only one vendor will complete the project. This approach has fostered a wider range of concepts that will be available for selection in the final design, construction, and operation.

RECOMMENDATIONS

Based on a careful assessment of the information presented to the RP and the findings developed in response to the review criteria, the RP provides the following recommendations:

1. The Project Team is urged to proceed with the implementation of its plans. In addition, the Project Team should also consider Recommendations 2 and 3.
2. The Project Team should consider additional testing for CSSX using real waste for extended periods of time, preferentially coupled with contactor operation.
3. The Project Team should investigate the attrition of the loaded MST during the slurry transfer to the DWPF to assess safety concerns.

Supplemental Information on MST Attrition, Settling, and Transfer Characteristics

Documents provided to the Review Panel and presentations by the Project Team (PT) during the peer review meeting did not include specific information or data on an issue of concern dealing with monosodium titanate (MST). The issue of concern is the attrition of MST which could lead to its accumulation in waste transfer lines. A question on this issue was raised by a Review Panel member during the presentations, and the Project Team had no information readily available to present in response to the question. Thus, the Project Team is providing this supplementary information in response to that question.

MST Particle Size Data

Poirier et al. (2003a) reviewed and compiled typical data from prior measurements of the size of simulated sludge, actual sludge, and (MST) particles. For the actual waste, all available data from prior measurements was collected. Since few prior measurements exist and since these analyses occurred using different analytical methods that span more than two decades, the PT cannot verify the consistency of the methods used to make the measurements.

Table 10 and Figs. 32 and 33 show the particle size of six vendor qualification MST samples. The samples are from drums of MST slurry manufactured for use at the In-Tank Precipitation (ITP) Facility in the mid-1990s. One sample came from Boulder Scientific and five samples came from Optima. The MST from Boulder Scientific was used in rotary microfilter testing. The MST from Optima was qualified for use at the Actinide Removal Process. The Optima samples are from a single lot.

The review found the procured MST samples—retrieved from 55-gallon (208 L) drums—to have a median particle size of 4-5 μm; a maximum particle size of 16 μm; and approximately 2 vol % of the particles smaller than 1 μm. There is a good agreement among five Optima samples.

Current and past procurement specifications require that the vendor provide MST as a slurry with no more than 1 vol % of the particles smaller than 1 μm and no more than 1 vol % of the particles larger than 35 μm. Both attrition and clumping can occur during storage. The clumping can become significant when storage occurs without temperature control.

Table 10. MST Particle Size.

Size (µm)	BSC-265-107 (vol %)	Optima 00-QAB-417#1 (vol %)	Optima 00-QAB-417#2 (vol %)	Optima 00-QAB-417#3 (vol %)	Optima 00-QAB-417#4 (vol %)	Optima 00-QAB-417#5 (vol %)
704	0	0	0	0	0	0
497.8	0	0	0	0	0	0
352	0	0	0	0	0	0
248.9	0	0	0	0	0	0
176	0	0	0	0	0	0
124.5	0	0	0	0	0	0
88	0	0	0	0	0	0
62.23	0	0	0	0	0	0
44	0	0	0	0	0	0
31.11	0	0	0	0	0	0
22	0	0	0	0	0	0
15.56	1.3	0.24	0.24	0.24	0.24	0.24
11	11.98	6.18	6.17	6.28	6.27	6.33
7.778	27.06	21.05	21.04	21.19	21.01	21.29
5.5	28.09	32.03	32.06	32.07	31.92	32.3
3.889	11.08	16.48	16.43	16.53	16.6	16.28
2.75	8.09	9.38	9.34	9.28	9.44	9
1.945	6.24	7.39	7.42	7.29	7.35	7.32
1.375	4.47	5.25	5.28	5.16	5.2	5.22
0.972	1.69	2	2.02	1.96	1.97	2.01

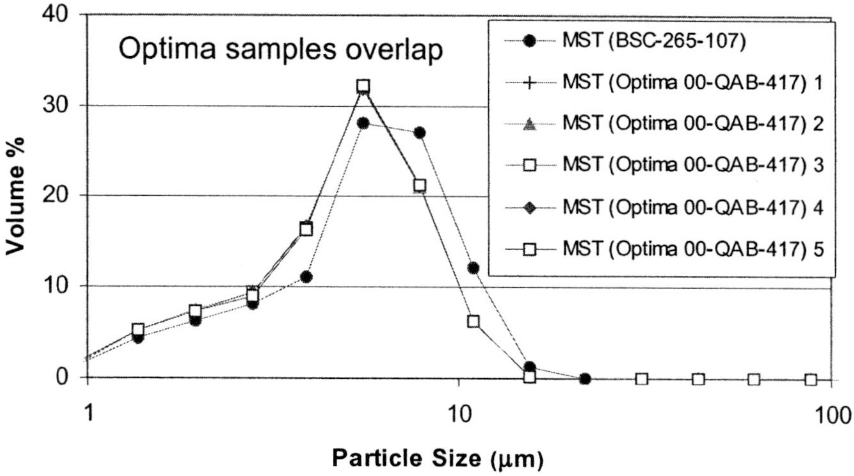

Fig. 32. MST Particle Size.

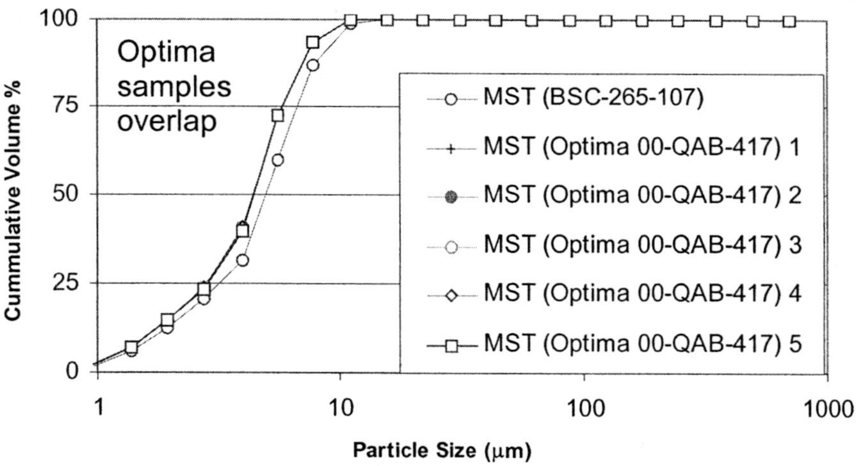

Fig. 33. MST Cumulative Particle Size.

Personnel generally need to use aggressive methods to re-suspend the settled solids. Hobbs (2003) showed that one drum of MST (Optima Batch #QAB-0-417) contained approximately 12 vol % of fines less than 0.9 μm (Fig. 34), apparently created during efforts to re-suspend the settled material.

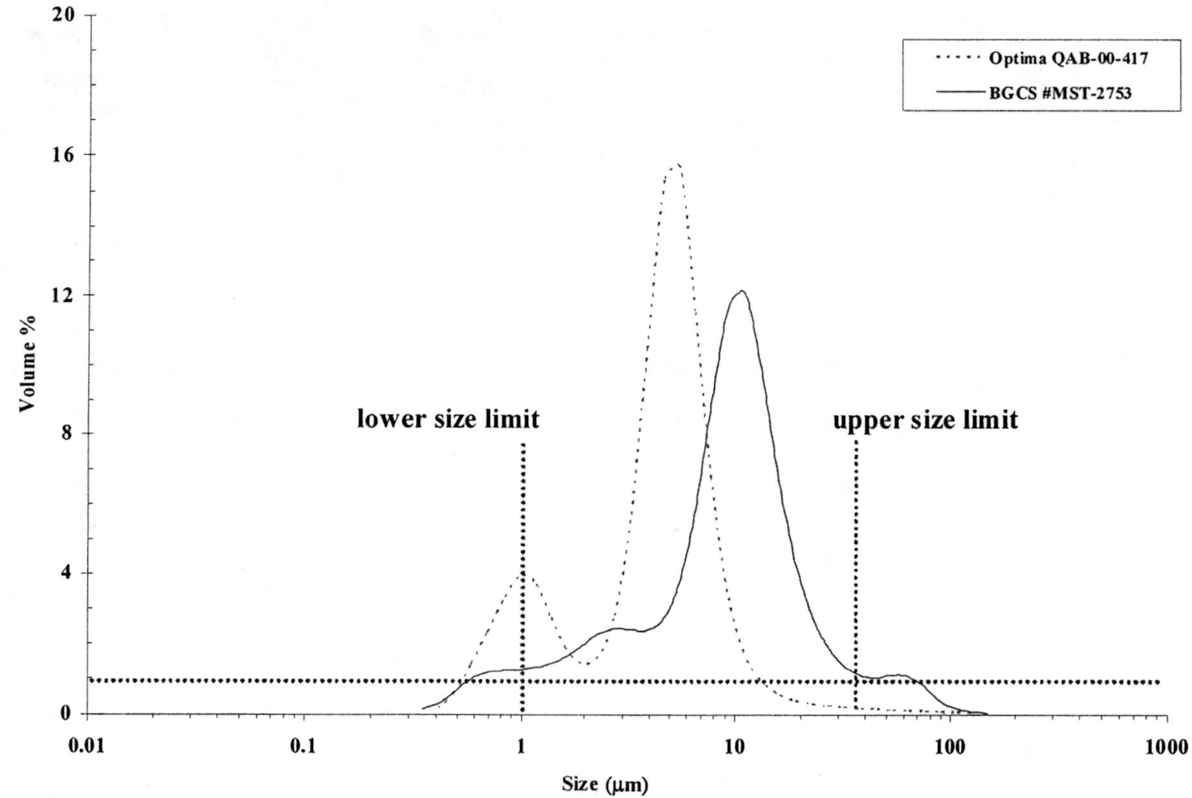

Fig. 34. Particle Size Distribution of MST Samples.

MST Attrition

Tests related to the concern of MST attrition were conducted during earlier development (Hobbs 1992) supporting the In-Tank Precipitation (ITP) Process. During the ITP process, breaking of the MST particles could have occurred as a result of high shear forces encountered in the slurry pumps and in the sintered metal filters. A decrease in the particle size may increase the adsorption capacity of the MST by increasing the number of sites available for adsorption. Therefore, experiments examined whether changes in the particle size affect the adsorption properties of the MST. The capacity of the MST to adsorb plutonium and uranium was measured to determine if the accumulation of a critical mass of fissile material could occur during the ITP process.

The particle size of the MST slurry was altered by physical separation and by mixing in a high-speed blender. Particle size distribution data for the "as-received" material; the two fractions obtained by the physical separation technique; and the slurry after various lengths of time at high speed in the blender are given in Table 11. While the particle size did change during high-speed blending, the modest change indicates substantial particle integrity. The physical separation experiment successfully removed the larger particles from the slurry producing a material, which was composed of a much higher fraction of smaller particles.

Particle size measurements during pilot-scale cross-flow filtration tests were made by Poirier et al. (2003b). Lasentec particle data indicated a small shift in particle sizes during the concentration range expected for Salt Waste Processing Facility (SWPF) (Fig. 35).

Table 11. Particle Size Distribution Data for MST.

MST Sample	Particle Size (μm)			
	1% <	50% <	99% <	Mean
"As-Received"	0.70	5.24	37.8	8.24
"large" particles	0.81	5.70	30.2	8.24
"small" particles	0.29	2.03	6.92	2.30
Blender 15 minutes	0.49	4.35	37.6	7.03
Blender 30 minutes	0.44	3.86	30.1	6.01
Blender 60 minutes	0.38	3.49	25.6	5.79
Blender 120 minutes	0.33	3.09	24.0	4.58

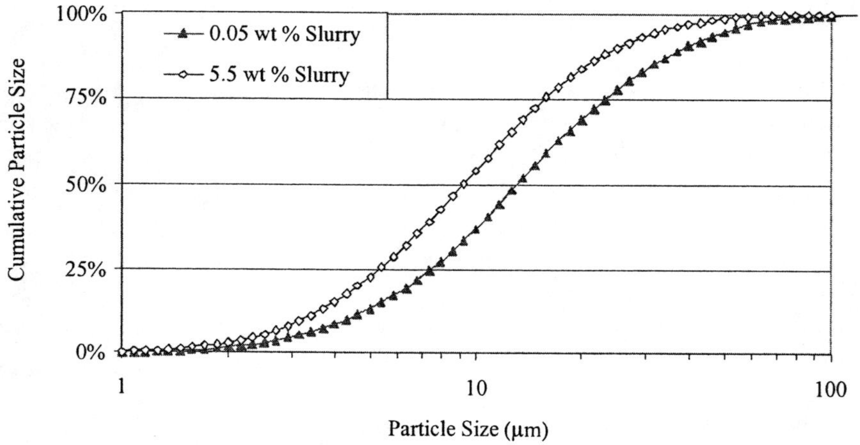

Fig. 35. Particle Size Data from 0.1 μm Filter Test at USC-FRED.

In conclusion, the particle size distribution of the MST may change by: 1) breaking due to high shear forces in pumps or in the cross-flow filter, or 2) separation due to differential settling in the tank. Smaller particles may exhibit higher adsorption due to increased surface area. However, experimental results indicated that changing the particle size distribution of the MST did not significantly affect the adsorption of uranium or strontium onto the MST.

Calculations on Slurry Transport in Horizontal Pipelines

The second aspect of this concern is whether formation of smaller particles by MST attrition could lead to accumulation in waste transfer lines and thus present a safety concern.

Solid-liquid horizontal flow can occur in a number of different flow regimes. The primary parameters influencing flow regimes are velocity and particle size. The common flow regimes are pseudo-homogeneous suspensions; heterogeneous suspensions; heterogeneous suspensions with sliding beds; and stationary beds. Pseudo-homogeneous suspensions occur at high velocities with small particles. The particles move at the same velocity as the fluid with a uniform distribution across the pipe. With slower velocities and larger particles, heterogeneous suspension occurs. The concentration of particles across the pipe is not uniform, and the particle velocity is slightly less than the fluid velocity. At low velocities with large particles, a heterogeneous suspension with a sliding bed occurs. Particles in the upper part of the pipe are in suspension and move with the liquid, while particles in the bottom of the pipe form a bed of solids that moves at a slower, uniform rate. At very low velocities with large particles, a stationary bed occurs. The upper part of the pipe contains a suspension, while the lower part contains a deposit, of which the surface layers move.

The conditions at which sliding and stationary beds occur are of interest because these conditions are normally undesirable. A sliding bed can cause substantial pipe abrasion. Sliding and stationary beds lead to low transport efficiencies. The transition between a heterogeneous suspension and a heterogeneous suspension with a sliding bed is often called the deposition velocity or re-suspension velocity, depending on whether the velocity is decreasing or increasing. The axial velocity in a transfer line should be greater than the deposition velocity or re-suspension velocity. Slurry transfers should occur as heterogeneous suspensions.

Theoretical calculations show that lower transport velocities are required to achieve a heterogeneous suspension with smaller particles (Lambert et al. 2003). Since the slurry is composed of a mixture of particle sizes, the fine particles that are easily suspended will reduce the settling rate of the coarse particles and thereby reduce the minimum transport velocity. Any particle attrition would make the slurry transport more efficient, eliminating the possibility of solids accumulation due to attrition.

Based on the calculations, flushing of the transfer line after transfer was recommended as some solids with larger particle sizes are expected to settle in the transfer line (Lambert et al. 2003).

Savannah River Site Slurry Transport Experience

Interarea transfer lines transport waste solutions between areas at the Savannah River Site (SRS). The three H-Area to S-Area lines, which transport waste solutions between H-Area Tank Farm and the Defense Waste Processing Facility (DWPF), each contain two 3-inch (7.5 cm) core pipes inside 10-inch (25 cm) jackets. Two of the pipes transport sludge; one transports recycle of low-level liquid waste from DWPF back to the Tank Farm; one was intended originally to transport ITP precipitate; and two are spares. The sludge lines transport sludge from Extended Sludge Processing (ESP) to the DWPF. The historical design basis for sludge transfers has been 4.3 ft/s (1.3 m/s) and 100 gpm (378 L/min). The present pumps produce 225 ft (68 m) of head at 100 gpm (378 L/min). The pumps must transfer sludge (13-19 wt % solids) from Tanks 40 and 51 to the Low Point Pump Pit (LPPP), a distance of approximately 3500 ft (1,050 m). After accounting for various pipe fittings, the effective length of piping is about 3650 ft (1,095 m). Transfers of MST/sludge from the Salt Waste Processing Facility (SWPF) to the LPPP (prior to DWPF) will be only several hundred feet depending on the final facility location.

The Project Team (PT) has carefully considered the potential of solids settling in transfer lines prior to and during the design phase of the SWPF project. The SRS has significant experience in evaluating and preventing solids from accumulating in transfer lines as a result of its almost 50 years of experience in transferring highly-radioactive liquid waste containing various types of solids. The SRS has not observed solids buildup in a forced transfer of high solids material even in frequently-used lines such as the F and H Interarea Transfer Line. Evaluations of material transfers are routinely conducted and documented prior to executing a transfer.

The PT has addressed this potential risk further by requiring the Engineering, Procurement, and Construction (EPC) contractors to design the liquid waste transfer system such that: 1) slurry transfers with high solids content are performed at sufficient velocity to minimize settling during transfer, and 2) any slurry transfer with high solids shall be followed by a flush of the transfer line with sufficient velocity and volume to remove essentially all settled solids. Implementation of these requirements has been verified in the conceptual designs of the EPCs.

References

Arafat, H.A. Solvent recovery from aqueous effluent using process diluent in the caustic side solvent extraction (CSSX) process, ANL-02/31. Chicago, IL: Argonne National Laboratory; (February 20) 2002.

Arafat, H.A.; Hash, M.C.; Hebden, A.S.; Leonard, R.A. Characterization and recovery of solvent entrained during the use of centrifugal contactors, ANL-02/08. Chicago, IL: Argonne National Laboratory; (October) 2001.

ASME (The American Society of Mechanical Engineers). Manual for peer review. Washington, DC: ASME; 2003.

ASME/RSI (The American Society of Mechanical Engineers/Institute for Regulatory Science). Assessment of technologies supported by the U.S. Department of Energy; results of the peer review for fiscal year 1997, CRTD Vol. 47. New York: ASME; 1997.

ASME/RSI (The American Society of Mechanical Engineers/Institute for Regulatory Science). Assessment of technologies supported by the U.S. Department of Energy; results of the peer review for fiscal year 1998, CRTD Vol. 50. New York: ASME; 1998.

ASME/RSI (The American Society of Mechanical Engineers/Institute for Regulatory Science). Assessment of technologies supported by the U.S. Department of Energy; results of the peer review for fiscal year 1999, CRTD Vol. 56. New York: ASME; 1999.

ASME/RSI (The American Society of Mechanical Engineers/Institute for Regulatory Science). Assessment of technologies supported by the U.S. Department of Energy; results of the peer review for fiscal year 2000, CRTD Vol. 61. New York: ASME; 2000.

ASME/RSI (The American Society of Mechanical Engineers/Institute for Regulatory Science). Strategy for remediation of groundwater contamination at the Nevada Test Site; technical peer review report; report of the review panel, CRTD Vol. 62. New York: ASME; 2001a.

ASME/RSI (The American Society of Mechanical Engineers/Institute for Regulatory Science). Requirements for disposal of remote-handled transuranic wastes at the waste isolation pilot plant; technical peer review report; report of the review panel, CRTD Vol. 63. New York: ASME; 2001b.

ASME/RSI (The American Society of Mechanical Engineers/Institute for Regulatory Science). Assessment of technologies supported by the U.S. Department of Energy; results of the peer review for fiscal year 2001, CRTD Vol. 64. New York: ASME; 2001c.

ASME/RSI (The American Society of Mechanical Engineers/Institute for Regulatory Science). Waste isolation pilot plant initial report for polychlorinated biphenyl disposal authorization; technical peer review report; report of the review panel, CRTD Vol. 65. New York: ASME; 2002a.

ASME/RSI (The American Society of Mechanical Engineers/Institute for Regulatory Science). Airborne release fractions; technical peer review report; report of the review panel, CRTD Vol. 68. New York: ASME; 2002b.

ASME/RSI (The American Society of Mechanical Engineers/Institute for Regulatory Science). The beryllium oxide manufacturing process; technical peer review report; report of the review panel, CRTD Vol. 69. New York: ASME; 2002c.

ASME/RSI (The American Society of Mechanical Engineers/Institute for Regulatory Science). Assessment of technologies supported by the U.S. Department of Energy; results of the peer review for fiscal year 2002; CRTD Vol. 70. New York: ASME; 2002d.

ASME/RSI (The American Society of Mechanical Engineers/Institute for Regulatory Science). Radionuclide transport in the environment; technical peer review report; report of the review panel; CRTD Vol. 71. New York: ASME; 2002e.

ASME/RSI (The American Society of Mechanical Engineers/Institute for Regulatory Science). Review of selected nuclear safety programs at Savannah River Site; technical peer review report; report of the review panel, CRTD Vol. 72. New York: ASME; 2003a.

ASME/RSI (The American Society of Mechanical Engineers/Institute for Regulatory Science). Hanford Site 100 B/C Risk Assessment Pilot Project; technical peer review report; report of the review panel, CRTD Vol. 73. New York: ASME; 2003b.

Barnes, M.J.; Edwards, T.B.; Hobbs, D.T.; Marshall, K.M. Strontium and actinides removal testing with monosodium titanate and other sorbents, WSRC-TR-2001-00436. Rev. 0. Aiken, SC: Westinghouse Savannah River Company; (October 29) 2001.

Birdwell, J.F.; Anderson, K.K. Evaluation of mass transfer performance for caustic-side solvent extraction of cesium in a conventional 5-cm centrifugal contactor, ORNL/TM-2001/278. Oak Ridge, TN: Oak Ridge National Laboratory; (January) 2002.

Birdwell, Jr., J.F.; Anderson, K.K.; Hobson, D.E. Investigation of emulsion formation in solvent treatment in the caustic-side solvent extraction (CSSX) process, ORNL/TM-2002/126. Oak Ridge, TN: Oak Ridge National Laboratory; (June) 2002.

Bonnesen, P.V.; Sloop, Jr., F.V.; Engle, N.L. Stability of the caustic-side solvent extraction (CSSX) process solvent: Effect of high nitrite on solvent nitration, ORNL/TM-2002/115. Oak Ridge, TN: Oak Ridge National Laboratory; (July) 2002.

Campbell, S.G. et al. Demonstration of caustic-side solvent extraction with Savannah River Site high level waste, WSRC-TR-2001-00223. Rev. 0. Aiken, SC: Westinghouse Savannah River Company; (April 19) 2001.

Campbell, J.A.; Hoppe, E.W.; Greenwood, L.R.; Farmer, III, O.T. Organic and actinide characterization of SRS tank samples, PNNL-14039. Draft. Richland, WA: Pacific Northwest National Laboratory; (September) 2002a.

Campbell, J.A.; Greenwood, L.R.; Farmer, III, O.T.; Hoppe, E.W. Technical task plan RL31WT21 subtask C, organic and actinide characterization, PNNLTP-43707-001. Rev. 1. Richland, WA: Pacific Northwest National Laboratory; (March 26) 2002b.

Delmau, L. H. et al. Caustic-side solvent extraction: Chemical and physical properties of the optimized solvent, ORNL/TM-2002/190. Oak Ridge, TN: Oak Ridge National Laboratory; (October) 2002a.

Delmau, L.H.; Haverlock, T.J.; Moyer, B.A. Caustic-side solvent extraction: Anti-caking surfactants found to be cause of apparent effect of high nitrite concentration on cesium stripping, ORNL/TM-2002/104. Oak Ridge, TN: Oak Ridge National Laboratory; (May) 2002b.

Delmau, L.H.; Bostick, D.A.; Haverlock, T.J.; Moyer, B.A. Caustic-side solvent extraction: Extended equilibrium modeling of cesium and potassium distribution behavior, ORNL/TM-2002/116. Oak Ridge, TN: Oak Ridge National Laboratory; (May) 2002c.

Dimenna, R.A. et al. Bases, assumptions, and results for the decision phase alternatives, WSRC-RP-99-00006. Rev. 3. Aiken, SC: Westinghouse Savannah River Company; (May) 2001.

DOE (U.S. Department of Energy). Savannah River Site salt processing alternatives. Final supplemental environmental impact statement, DOE/EIS-0082-S2. Aiken, SC: DOE Savannah River Operations Office; (June) 2001.

Fink, S.D.; Hobbs, D.T.; Poirier, M.R. Status assessment of readiness to deploy permanganate treatment for removal of strontium and actinides from Savannah River Site high level waste, WSRC-RP-2002-00524. Aiken, SC: Westinghouse Savannah River Company; (September 23) 2002a.

Fink, S.D.; Hobbs, D.T.; Peters, T.B. Demonstration of MST efficacy on removal of actinides and strontium in "bounding alpha" waste, WSRC-TR-2002-00555. Aiken, SC: Westinghouse Savannah River Company; (December 17) 2002b.

Harmon, H.; Schlahta, S.; Wester, D.; Walker, J.; Fink, S.; Thompson, M. Tanks focus area: Savannah River Site processing project. Research and development summary, TFA-0105. Rev. 0. Richland, WA: Pacific Northwest National Laboratory; (May) 2001a.

Harmon, H.; Murphy, J.; Smalley, R. Savannah River Site salt processing project down selection decision analysis summary report. TFA-0106. Rev. 0. Richland, WA. Pacific Northwest National Laboratory; (June) 2001b.

Harmon, H.; Leugemors, R.; Fink, S.; Thompson, M.; Walker, D. Tanks focus area: Summary report on caustic-side solvent extraction chemical and physical properties, TFA-0220. Rev. 0. Richland, WA: Pacific Northwest National Laboratory; (September) 2002a.

Harmon, H.; Leugemors, R.; Fink, S.; Thompson, M.; Walker, D. Tanks focus area: Summary report on caustic-side solvent extraction engineering tests of equipment, TFA-0221. Rev. 0. Richland, WA: Pacific Northwest National Laboratory; (September) 2002b.

Harmon, H.; Leugemors, R.; Fink, S.; Thompson, M.; Walker, D. Tanks focus area: Summary report on caustic-side solvent extraction process chemistry, TFA-0222. Rev. 0. Richland, WA: Pacific Northwest National Laboratory; (September) 2002c.

Harmon, H.; Leugemors, R.; Fink, S.; Thompson, M.; Walker, D. Tanks focus area: Summary report on caustic-side solvent extraction actual waste tests, TFA-0223. Rev. 0. Richland, WA: Pacific Northwest National Laboratory; (September) 2002d.

Herman, D.T.; Poirier, M.R.; Hobbs, D.T.; Fink, S.D. Testing of the SpinTek Rotary Microfilter Using Actual Waste, WSRC-TR-2003-00030. Rev. 0. Aiken, SC: Westinghouse Savannah River Company; (February) 2003.

Hobbs, D.T. Plutonium and uranium adsorption on monosodium titanate, WSRC-RP-92-093. Aiken, SC: Westinghouse Savannah River Company; (August 13) 1992.

Hobbs, D.T. Potential for the precipitation of uranium and plutonium solids upon addition of nitric acid to waste solutions in a caustic-side solvent extraction process, WSRC-TR-2002-00054. Rev. 0. Aiken, SC: Westinghouse Savannah River Company; (May 10) 2002.

Hobbs, D.T. Characterization of monosodium titanate sample supplied by blue grass chemical specialties, WSRC-RP-2003-01084. Aiken, SC: Westinghouse Savannah River Company; (November 20) 2003.

Klatt, L.N. et al. Caustic-side solvent extraction solvent-composition recommendations, ORNL/TM-2001/258. Oak Ridge, TN: Oak Ridge National Laboratory; (January) 2002.

Lambert, D.P.; Peters, T.B.; Stallings, M.E.; Poirier, M.R.; Fink, S.D. Determining the cause of low flowrates during Am/Cm simulant testing in F area, WSRC-TR-2002-00569. Rev. 0. Aiken, SC: Westinghouse Savannah River Company; (April 29) 2003.

Law, J.D.; Tillotson, R.D.; Todd, T.A. Evaluation of the hydraulic performance and mass transfer efficiency of the CSSX process with the optimized solvent in a single stage of 5.5-cm diameter centrifugal contactor, INEEL/EXT-02-01109. Idaho Falls, ID: Idaho National Engineering and Environmental Laboratory; (September) 2002.

Lee, D.D.; Birdwell, Jr., J.F.; Bonnesen, P.V.; Tompkins, B.A. Density changes in the optimized CSSX solvent system, ORNL/TM-2002/204. Oak Ridge, TN: Oak Ridge National Laboratory; (November) 2002.

Leonard, R.A. Caustic-side solvent extraction flowsheet for optimized solvent, ANL-02/19. Chicago, IL: Argonne National Laboratory; (October) 2001.

Leonard, R.A.; Regalbutto, M.C.; Aase, S.B.; Arafat, H.A.; Falkenberg, J.R. Hydraulic performance of a 5-cm CINC contactor for caustic-side solvent extraction, ANL-02/18. Chicago, IL: Argonne National Laboratory; (December 12) 2001a.

Leonard, R.A. et al. Interim report on a multiday test of the caustic-side solvent extraction flowsheet for cesium removal from a simulated SRS tank waste, ANL-01/10. Chicago, IL: Argonne National Laboratory; (April) 2001b.

Leonard, R.A. et al. Simulant flowsheet test with modified solvent for cesium removal using caustic-side solvent extraction, ANL-02/22. Chicago, IL: Argonne National Laboratory; (April 22) 2002.

Maskarinec, M.P.; Walker, J.F.; Moyer, B.A. Analysis of a sample of the solvent used in the Argonne National Laboratory multiday flowsheet test, CERS/SR/SX/027. Oak Ridge, TN: ORNL Chemical and Energy Research Section; (February 19) 2002.

Moyer, B.A. et al. Caustic-side solvent extraction chemical and physical properties, progress in FY 2000 and FY 2001, ORNL/TM-2001/285. Oak Ridge, TN: Oak Ridge National Laboratory; (February) 2002.

Norato, M.A. et al. High level waste demonstration of the caustic-side solvent extraction process with optimized solvent in the 2-cm centrifugal contactor apparatus, WSRC-TR-2002-00243. Draft. Aiken, SC: Westinghouse Savannah River Company; (August 2) 2002a.

Norato, M.A.; Fink, S.D.; Fondeur, F.F.; Kessinger, G.F.; Pierce, R.A.; Walker, D.D. Demonstration of the caustic-side solvent extraction process in the 2-cm centrifugal contactor apparatus using dissolved salt cake from tank 37H. WSRC-TR-2002-00307. Rev. 0. Aiken, SC: Westinghouse Savannah River Company; (August 23) 2002b.

NRC (National Research Council). Alternatives for the high-level waste salt processing at the Savannah River Site. Washington, DC: National Academy Press; 2000.

NRC (National Research Council). Evaluation for selecting a salt processing alternative for high-level waste at the Savannah River Site: Interim Report. Washington, DC: National Academy Press; 2001a.

NRC (National Research Council). Research and development on a salt processing alternative for high-level waste at the Savannah River Site. Washington, DC: National Academy Press; 2001b.

Pennebaker, F.M.; DiPrete, D.P.; Malek, M.A. Analytical methods for Cs-137 and other radionuclides in solvent samples, WSRC-TR-2002-308. Aiken, SC: Westinghouse Savannah River Company; (July 17) 2002.

Pereira, C.; Arafat, H.A.; Falkenberg, J.R.; Regalbuto, M.C.; Vandegrift, G.F. Recovery of entrained CSSX solvent from caustic aqueous raffinate using coalescers, ANL/CMT/CSSX/2002/09. Chicago, IL: Argonne National Laboratory; (September) 2002.

Peters, T.B.; Fink, S.D.; Hobbs, D.T.; Norato, M.A.; Walker, D.D. Demonstration of MST and permanganate efficiency on removal of actinides and strontium from Savannah River Site high level waste, WSRC-TR-2002-00355. Aiken, SC: Westinghouse Savannah River Company; (March 11) 2003.

Poirier, M.R.; Siler, J.L.; Fink, S.D. Filtration of actual Savannah River waste treated with permanganate or monosodium titanate, WSRC-TR-2002-00134. Rev. 0. Aiken, SC: Westinghouse Savannah River Company; (March 14) 2002.

Poirier, M.R.; Fink, S.D. Recommendation for using smaller (0.1μ) pore-size media for filtration in salt waste processing project, WSRC-TR-2002-00341. Aiken, SC: Westinghouse Savannah River Company; (August 5) 2002.

Poirier, M.R.; Fondeur, F.F.; Lambert, D.P.; Hobbs, D.T.; Fink, S.D. Particle size of simulated SRS sludge, actual SRS sludge, and monosodium titanate, WSRC-TR-2003-00221. Aiken, SC: Westinghouse Savannah River Company; (May 20) 2003a.

Poirier, M.R.; Siler, J.L.; Fink, S.D. Pilot-scale testing of a 0.1 micron filter with SRS simulated high level waste, WSRC-TR-2003-00469. Rev. 0. Westinghouse Savannah River Company; (October 8) 2003b.

RSI (Institute for Regulatory Science). Handbook of peer review. Columbia, MD: RSI; 2003.

SROO (Savannah River Operations Office). Inside Savannah River, July 2002. Available at http://sro.srs.gov/inside1.htm

SRS (Savannah River Site). Historical highlights @ SRS, April 2002. Available at http://www.srs.gov/general/about/hilits.htm

Steele, W.V. Vapor pressure of 1-2,2,3,3-tetrafluoropropoxy)-3-(4-secbutylpenoxy)-2-propanol, ORNL/TM-2001/287. Oak Ridge, TN: Oak Ridge National Laboratory; (January) 2002.

Stock, L.M.; Camaioni, D.M.;Wester, D.W.; Gutsche, C.D. Decomposition pathways for the Savannah River Site caustic-side solvent extraction system, PNNL-13935. Oak Ridge, TN: Oak Ridge National Laboratory; (July) 2002.

USNRC (U.S. Nuclear Regulatory Commission). Licensing requirements for land disposal of radioactive waste, 10 CFR 61.55. Washington, DC: Government Printing Press; 2003.

Walker, D.D. Stability and solubility tests with SRS simulated wastes, WSRC-TR-2000-00273. Rev. 0. Aiken, SC: Westinghouse Savannah River Company; (November 6) 2000.

Walker, D.D. Stability tests with actual Savannah River Site radioactive waste, WSRC-TR-2002-00026. Rev. 0. Aiken, SC: Westinghouse Savannah River Company; (July 15) 2002a.

Walker, D.D. Organic Compounds in Savannah River Site High-Level Waste. WSRC-TR-2002-00391. Draft A. Aiken, SC: Westinghouse Savannah River Company;(August 30) 2002b.

Walker, D.D.; Beasley, M.H.; Coleman, A.D.; Croy, B.H.; DiPrete, C.C.; DiPrete, D.P. Solvent extraction batch distribution coefficients with SRS dissolved salt cake, WSRC-TR-2001-00533. Rev. 1. Aiken, SC: Westinghouse Savannah River Company; (March 26) 2002.

Wilmarth, W.R.; Fink, S.D.; Hobbs, D.T.; Hay, M.S. Characterization and dissolution studies of samples from the 242-16H evaporator gravity drain line, WSRC-TR-97-00326. Rev. 0. Aiken, SC: Westinghouse Savannah River Company; (October 16) 1997.

Wilmarth, W.R. et al. Characterization and dissolution studies of samples from the 242-16H evaporator wall, WSRC-TR-2000-00038. Rev. 0. Aiken, SC: Westinghouse Savannah River Company; (January 31) 2000a.

Wilmarth, W.R.; Coleman, C.J.; Hart, J.C.; Boyce, W.T. Characterization of samples from the 242-16H evaporator wall, WSRC-TR-2000-00089. Rev. 0. Aiken, SC: Westinghouse Savannah River Company; (March 20) 2000b.

Wilmarth, W.R.; Dukes, V.H.; Mills, J.T.; DiPrete, C.C.; DiPrete, D.P. Solvent composition impacts on the extraction of uranium and plutonium in the caustic-side solvent extraction process, WSRC-TR-2002-00060. Aiken, SC: Westinghouse Savannah River Company; (January 18) 2002a.

Wilmarth, W.R. et al. Caustic-side solvent extraction batch distribution measurements for SRS high level waste samples and dissolved salt cake, WSRC-TR-2002-00336. Rev. 0. Aiken, SC: Westinghouse Savannah River Company; (July 19) 2002b.

WSRC (Westinghouse Savannah River Company). Savannah River Site High Level Waste Salt Disposition Systems Engineering Team, Final Report: Recommendation Pre-Conceptual Design, WSRC-RP-98-00170. Rev. 0. Westinghouse Savannah River Company; (October 29) 1998.

WSRC (Westinghouse Savannah River Company). Salt waste processing facility actinides/Sr removal research and development reports. Aiken, SC: Westinghouse Savannah River Company; 2003a.

WSRC (Westinghouse Savannah River Company). Salt waste processing facility CSSX research and development reports. Aiken, SC: Westinghouse Savannah River Company; 2003b.

Biographical Summaries

Jimmy T. Bell is currently President of Bell Consultants. With over three decades of professional experience, his principal areas of expertise include: 1) management and planning of separations technologies and chemical flowsheets; 2) development of separation technologies in solvent extraction, ion exchange, liquids-solids separations, and isotope enrichment technologies; 3) nuclear fuels technologies—specifically, fuels preparation, mixed oxide preparations, severe accident source terms, plutonium chemistry, tritium technologies, and fuels reprocessing; and 4) waste management, including waste minimization, nuclear waste separations, and mixed waste technologies. Previously, he served as Head of the Chemical Development Section at Oak Ridge National Laboratory, managing several programs including "Comprehensive Treatment of U.S. Department of Energy Tank Wastes"; "Molten Salt Oxidation Treatment of Mixed Wastes"; "Fission Product Release and Behavior in Severe LWR Accidents"; "Atomic Vapor Laser Isotope Separator Feed and Product Processing"; "Nuclear Wastes Separations Technologies"; "Thermodynamics and Stabilities of Superconducting Materials"; "Nuclear Fuel Reprocessing Technologies"; and "Purification of Uranium Alloys". He has served as Chair and member of several panels for U.S. Department of Energy facilities. Jimmy T. Bell is a member of the American Chemical Society, the American Nuclear Society, and the American Institute of Chemical Engineers. He was Associate Editor of *Separation Science and Technology*. He has authored or coauthored over 45 papers and publications. He received a B.S. degree in Chemistry from Berry College, and a Ph.D. in Physical Chemistry from the University of Mississippi.

Gary A. Benda is President of U.S. Energy Corp.—an environmental management firm specializing in radioactive mixed waste management, health physics, decontamination and decommissioning, and technology development. Previously, he was Vice-President, General Manager of the Programs Division for NUKEM Nuclear Technologies, Inc. His responsibilities included developing and maintaining federal programs in North America that specialized in engineering and waste-processing services. Prior to NUKEM, he spent over 17 years with Chem-Nuclear Systems/WMX Technologies in various management roles. He also directed the site investigation, geophysical analysis, site screening, and license application, as well as managed the public hearings and licensing operations associated with local and national regulatory agencies for new low-level waste sites. He has over 20 years of experience in environmental restoration, technology development, and waste management, and has instructed over 20 national and international professional courses on radioactive waste management, mixed waste, and technology development. He is a member of the American Society of Mechanical Engineers (ASME), American Nuclear Society, and Health Physics Society. He has served as Chair of the ASME National Mixed Waste Committee, Environmental Remediation Committee, and Environmental Engineering Division. He has also chaired over 100 technical sessions at numerous national and international conferences on environmental management. He has authored and coauthored various scientific papers, reports, book chapters, and articles on the nuclear environment. Gary Benda is a Certified Health Physicist. He received a B.S. in Health Physics from Oklahoma State University, an M.S. in Applied Nuclear Science from Georgia Institute of Technology, and an M.B.A. from Seattle City University.

Erich W. Bretthauer is a consultant. Previously, he held the position of research professor at the University of Nevada-Las Vegas from January 1993 to March 1995. In that capacity, he served as Executive Director of Nevada Industry, Science Engineering & Technology, a public-private partnership which developed programs to enhance the scientific infrastructure of the state of Nevada. He was also the Assistant Administrator for Research and Development at the U.S. Environmental Protection Agency (EPA) from March 1990 until January 1993. In that capacity, he managed the Research and Development activities of a large and multi-disciplinary agency. Erich Bretthauer rose through the ranks of the EPA and served in a number of capacities ranging from a bench scientist to policy manager at national and international levels. He directed the EPA's emergency and long-term monitoring program after the accident at Three Mile Island, as well as its bioremediation program in Prince William Sound after the Valdez oil spill. Erich Bretthauer was the leader of the U.S. delegation and co-leader of a five-year North Atlantic Treaty Organization project which focused on exposures, risks, and measures to control dioxins. He also directed the EPA's ecological research program, and was Director of EPA's Environmental Monitoring Systems Laboratory in Las Vegas. He is a member of Sigma Xi; the American Chemical Society; and the American Water Works Association; and has served on the Federal Advisory Committee to the Civil Engineering Research Foundation. Erich Bretthauer is the author and coauthor of numerous papers, reports, and other publications. He received his B.S. and M.S. in chemistry from the University of Nevada, Reno, NV.

Ernest L. Daman is Chairman Emeritus of Foster Wheeler Development Corporation where he previously served as Director of Research and Chairman of the Board. He also held the position of Senior Vice President at the parent company, FWC. He is a Past President of American Society of Mechanical Engineers and was elected to the National Academy of Engineering. Ernest Daman is a Fellow of the Institute of Energy (England) and the American Association for the Advancement of Science, and Past Chairman of the American Association of Engineering Societies. He served on several American Society of Mechanical Engineers committees as member or chairman. Ernest Daman is the author of numerous papers and holds 18 patents. He was responsible for the design and development of a combined steam gas turbine plant, fluidized bed combustion, fast breeder reactor components, supercritical steam generators, environmental control processes, and advanced high-efficiency power generation systems. Ernest Daman received his B.M.E. degree from the Polytechnic Institute of Brooklyn.

Irwin Feller is Director of the Institute for Policy Research and Evaluation and Professor of Economics at The Pennsylvania State University, where he has been on the faculty since 1963. His current research interests include the economics of academic research, the University's role in technology-based economic development, and the evaluation of federal and state technology programs. He is the author of *Universities and State Governments*: A Study in Policy Analysis, and over 100 refereed journal articles, final research reports, book chapters, reviews, and numerous papers presented to academic, professional, and private organizations. He is former Chair of the Committee on Science, Engineering, and Public Policy, American Association for the

Advancement of Science. Irwin Feller was the American Society of Mechanical Engineers Pennsylvania State Fellow in 1996-1997. He has been appointed to the National Research Council's Committee on Science, Engineering, and Public Policy; International Benchmarking of U.S. International Competitiveness-Immunology; Transportation Research Board, Research and Technology Coordinating Committee, National Research Council; and National Institute of Standards and Technology-Manufacturing Extension Partnership National Advisory Board. Irwin Feller is Chair of the National Science Foundation's Advisory Committee on Social, Behavioral, and Economic Sciences. He received a B.B.A. in Economics from the City University of New York and a Ph.D. in Economics from the University of Minnesota.

Robert A. Fjeld is Dempsey Professor of Environmental Engineering and Science at Clemson University. He coordinates the Department's nuclear environmental focus area, which is concerned with the environmental aspects of nuclear technologies including health physics, radioactive waste management, and risk assessment. Previously, he served as a faculty member in the Nuclear Engineering Department at Texas A&M University. He has active research on actinide transport in soils, instrumentation for measuring radioactivity in environmental samples, and environmental risk assessment. Robert Fjeld is a member of the Health Physics Society, American Nuclear Society, Society for Risk Analysis, and the American Society of Mechanical Engineers, where he serves as newsletter editor for the Mixed Waste Committee. He has served on two NRC Committees studying decontamination and decommissioning issues. Robert Fjeld is the author or coauthor of over 80 technical publications and presentations on topics such as radiation measurements, environmental transport of radionuclides, risk assessment, and aerosol physics. Robert Fjeld received a B.S. degree from North Carolina State University; and an M.S. degree and a Ph.D. from The Pennsylvania State University, all in Nuclear Engineering. He is a registered Professional Engineer.

William T. Gregory, III is currently Principal of Vinculum Marketing Solutions. Prior to forming Vinculum, he was Director of Government Programs for Foster Wheeler Environmental Corporation, an engineering and construction firm providing environmental and waste management services to government and private sector clients worldwide. Previously, he held a number of operational and business development positions at equipment manufacturing and service provision firms supporting nuclear utilities, industrial and process industries, and government agencies. His work has involved the management, processing, and disposition of hazardous, radioactive, and mixed wastes. He has also worked on the decontamination and decommissioning of nuclear facilities and on providing a wide range of environmental services in response to regulatory drivers. Prior to entering the private sector, he served with the U.S. Navy on nuclear submarines and at the operational command center for submarine operations in the Atlantic Fleet. William Gregory is actively involved with a number of international, national, and local organizations including the American Society of Mechanical Engineers and the American Nuclear Society. He is a founding member of the Board of Directors for the annual international Waste Management Symposium. William Gregory has served as an elected officer of several American Society of Mechanical Engineers divisions. He received a B.S. degree in Geology from the University of New Mexico, and an M.B.A. degree from Lamar University. He also attended naval nuclear power, nuclear weapons, and engineering schools as a U.S. Navy officer.

Nathan H. Hurt is a consultant in management and engineering with Technical and Management Consulting. He provides services to industrial firms and government agencies involved in environmental clean-up and waste management—both chemical and radioactive. He has extensive experience in the areas of executive management; plant management; engineering management; project management; marketing; and sales. He specializes in the areas of: uranium enrichment/production; engineering; development and marketing; plant management of rubber chemicals; petrochemicals; and thermoplastics. He also specializes in the engineering management of synthetic rubber and lattices; vinyl monomers and copolymers; polyesters; U.S. Department of Energy (DOE) weapons plants; quality assurance management; and operational readiness review. Nathan Hurt has been involved with the decommissioning of nuclear facilities. He was the Corporate Sponsor or Program Manager for seven decommissioning contracts at the DOE Complexes in Oak Ridge, TN;

and Pinellas, FL. Previously, Nathan Hurt worked for Sharp and Associates, Inc. as the Director and Project Manager at the Oak Ridge Office. He was Vice President and Director of Oak Ridge Operations for IDM Environmental Corp., where he was responsible for the marketing and sales of decontamination, decommissioning, and waste management. He served as Project Manager for the laboratory quality assurance program at Westinghouse Hanford; DOE's Rocky Flats Plant—plant-wide identification of electrical equipment. He managed a study for a waste treatment and storage facility at the Portsmouth Area Uranium Enrichment Facility which included incineration and compaction of low-level radioactive wastes. He also worked for The Goodyear Tire and Rubber Company, including Goodyear Atomic, as Director of Research and Development, and President, where he was responsible for the operation of the Portsmouth Area Uranium Enrichment Facility. Nathan Hurt is a Past President of the American Society of Mechanical Engineers. He has been a member of: the American Association of Engineering Societies' Board of Governors; the American Institute of Chemical Engineers; and the Institute of Nuclear Materials Management. He is also a member of Tau Beta Pi Honorary Engineering Society; Pi Tau Sigma Honorary Mechanical Engineering Society; and was a member of The Nuclear Engineering Advisory Board of Worcester Polytechnic Institute. Nathan Hurt received a B.S. degree in Mechanical Engineering from the University of Colorado and has done Graduate, Technical, and Management course work at Pennsylvania State University. He is a registered Professional Engineer in Ohio.

Michael C. (M.C.) Kirkland is Vice President for the Southeastern Region of the Institute for Regulatory Science (RSI). In that capacity he leads various RSI projects related to the RSI mission in the south-eastern U.S. Previously he was an independent consultant involved in peer review and various independent studies. For example, he led a team that performed an External Independent Review of the $1.3 billion Spallation Neutron Source Project at Oak Ridge, TN. He assisted in the planning and review of a management assessment at a U.S. Department of Energy (DOE) Site that involved the restart of a plutonium facility. He participated in planning, procurement, and review activities in the environmental remediation area that included decommissioning activities at a shut down nuclear test reactor; and designed and installed a ground water cleanup technology. M.C. Kirkland managed several environmental and construction projects that employed many soil investigative techniques including significant work with cone penetrometers. Additionally, he provided consulting services to a large environmental remediation services company regarding Dense Non-Aqueous Phase Liquid locating and removal techniques. During his tenure at the Savannah River Site (SRS) of DOE, M.C. Kirkland was a Technical Advisor, Project Manager, and Director of the Project Engineering Division. He evaluated nuclear and mixed waste conditions and aspects of high level wastes and spent nuclear fuel; determined material inventories; performed pollution prevention and environmental health and safety evaluations for a proposed waste treatment facility; served as technical advisor to a study administered by the Savannah River Operations Office; and developed integrated schedules defined for this project. M.C. Kirkland was director of the Project Engineering Division and managed the SRS design and construction program. He has been involved with waste management and environmental projects; cutting-edge technology programs; and worked with lasers and magnetic containment. He served as Director of the Waste and Fuel Cycle Technology Office, and planned and coordinated the programs of the DOE National High Level Waste Technology Office; the SR Fuel Cycle Technology Program; and the Commercial Interim Spent Fuel Management Program. M.C. Kirkland holds a B.S. in Mechanical Engineering from the University of South Carolina. He is registered as a Professional Engineer in South Carolina.

Peter B. Lederman is a consultant with over 48 years of experience in all facets of process engineering, environmental management, control, and policy development. This includes hazardous substance management; environmental remediation; environmental audit; pollution prevention; development of air pollution control devices; and reuse of waste products. He recently retired as Executive Director of the Center for Environmental Engineering & Science, Executive Director for Patents and Licensing, and Research Professor of Chemical Engineering and Environmental Policy at the New Jersey Institute of Technology. Peter Lederman managed major programs in industrial waste treatment research and development, and in oil and hazardous material spill control and remediation. Most recently, he was responsible for a study of the Economic Impact of Environmental Regulations. He has been responsible for technology transfer efforts including the maturing and licensing of innovative environmental technologies. He is a Fellow of the American Institute of Chemical Engineers; a Diplomat of the American Academy of Environmental Engineers; and a member of the ASME. He has served on several committees of the NRC and is the chair of the NRC Committee on Review and Evaluation of the Army Chemical Stockpile Disposal Program. He chaired the American Institute of Chemical Engineer's Environmental Division and is currently chair of its Societal Impacts Operating Council. Peter Lederman received a B.S.E., M.S.E., and Ph.D. (all in Chemical Engineering) from the University of Michigan in Ann Arbor, MI and is a registered Professional Engineer.

Betty R. Love is currently Executive Vice President of the Institute for Regulatory Science. In that capacity, she is responsible for the management of day-to-day operations of the Institute, and for administration of several projects. She is the Administrative Manager of a large-scale peer review program in collaboration with the American Society of Mechanical Engineers for a number of organizations including the U.S. Department of Energy. Her current research activities center around the development and implementation of a systematic approach to stakeholder participation, notably in scientific meetings. Previously, Betty Love was Director, Department of Training and Information within the Office of Environmental Health and Safety of Temple University in Philadelphia, PA. During that period she was instrumental in the development of a "Handbook of Environmental Health and Safety". She also developed and implemented a large-scale training program not only for the faculty and staff of the University but also for others. Betty Love is currently Managing Editor of *Technology*. She has published several papers in peer-reviewed journals; has edited a number of compendia; and is the primary author of *Manual for Public and Stakeholder Participation*. Betty Love received a B.S. degree in Business Administration from Virginia State University in Petersburg, VA, and an M.S. degree in Developmental Clinical Psychology from Antioch College in Yellow Springs, OH.

Jeffrey A. Marqusee is currently the Technical Director of the Strategic Environmental Research and Development Program (SERDP), and the Director of the Environmental Security Technology Certification Program (ESTCP). SERDP is a tri-agency (U.S. Department of Defense [DOD], U.S. Department of Energy, and U.S. Environmental Protection Agency) environmental research and development program managed by the DOD. SERDP supports research and development to solve environmental issues of relevance to DoD in the areas of cleanup, compliance, conservation and pollution prevention. ESTCP is a DOD-wide program designed to demonstrate innovative environmental technologies at DoD facilities. ESTCP provides for rigorous validation of the cost and performance of new environmental technologies in cooperation with the regulatory and end-user communities. Prior to his current position, Jeffrey Marqusee served as a program manager for environmental technology in the Office of the Deputy Under Secretary of Defense for Environmental Security. He was the principal advisor to the Deputy Under Secretary on environmental technology issues. Before joining DoD, he worked at the Institute for Defense Analyses, where he advised both DOD and National Aeronautics and Space Administration in the areas of remote sensing, environmental matters and military surveillance. Jeffrey Marqusee has worked at Stanford University, the University of California and the National Institute of Standards and Technology. He has a Ph.D. in Physical Chemistry from the Massachusetts Institute of Technology.

A. Alan Moghissi is currently President of the Institute for Regulatory Science (RSI), a non-profit organization dedicated to the idea that societal decisions must be based on best available scientific information. The activities of the Institute include research, scientific assessment, and science education at all levels—particularly the education of minorities. Previously, Alan Moghissi was Associate Vice President for Environmental Health and Safety at Temple University in Philadelphia, PA and Assistant Vice President for Environmental Health and Safety at the University of Maryland at Baltimore. In both positions, he established an environmental health and safety program and resolved a number of relevant existing problems in those institutions. As a charter member of the U.S. Environmental Protection Agency (EPA), he served in a number of capacities, including Director of the Bioenvironmental/Radiological

Research Division; Principal Science Advisor for Radiation and Hazardous Materials; and Manager of the Health and Environmental Risk Analysis Program. Alan Moghissi has been affiliated with a number of universities. He was a visiting professor at Georgia Tech and the University of Virginia, and was also affiliated with the University of Nevada and the Catholic University of America. Alan Moghissi's research has dealt with diverse subjects ranging from measurement of pollutants to biological effects of environmental agents. A major segment of his research has been on scientific information upon which laws, regulations, and judicial decisions are based—notably risk assessment. He has published nearly 400 papers, including several books. He is the Editor-in-Chief of *Technology: A Journal of Science Serving Legislative, Regulatory, and Judicial Systems*, which traces its roots to the *Journal of the Franklin Institute*—one of America's oldest continuously published journals of science and technology. Alan Moghissi is a member of the editorial board of several other scientific journals and is active in a number of civic, academic, and scientific organizations. He has served on a number of national and international committees and panels. He is a member of a number of professional societies including the American Society of Mechanical Engineers and is past chair of its Environmental Engineering Division. He is also an academic councilor of the Russian Academy of Engineering. Alan Moghissi received his education at the University of Zurich, Switzerland, and Technical University of Karlsruhe in Germany, where he received a doctorate degree in physical chemistry.

Lawrence C. Mohr, Jr., is currently Professor of Medicine, Biometry, and Epidemiology; and Director of the Environmental Biosciences Program at the Medical University of South Carolina. His areas of research and special interest include internal medicine and pulmonary disease—specifically diseases of the chest and respiratory system. An area of particular interest to Lawrence Mohr is environmental medicine, including molecular epidemiology and biomarker applications. He has been involved in studies related to environmental lung disease; patho-physiology; prevention and treatment of high altitude illness; high altitude physiology; risk assessment of environmental hazards and clinical epidemiology. Other areas of considerate interest to Lawrence Mohr are assessment of clinical outcomes; health policy analysis; and international health. This latter area includes: global epidemiology; medical relief operations; and health care in Central and Eastern Europe, as well as medical history—the impact of illness on world leaders. Previously, he held academic appointments as a Teaching Fellow in Medicine at the Uniformed Services University of the Health Sciences in Bethesda, MD. He was Associate Clinical Professor of Medicine and Emergency Medicine at George Washington University, Washington, DC. While in these institutions, he was a staff member of the Medical Support Group for the President of the United States. Lawrence Mohr was on the Medical Staff of Walter Reed Army Medical Center—where he completed his Internship and Residency in Internal Medicine—as well as George Washington University Hospital, both in Washington, DC. He has held Visiting Professorships at various universities. He served as Visiting Chief Resident at Presbyterian Hospital and Visiting Professor at the School of Nursing, both at Columbia University. Additionally, Lawrence Mohr was Visiting Professor of: William Beaumont Army Medical Center, Tulane University, University of Cincinnati, New York University, Brown University, East Carolina University, and the Mayo Clinic. Lawrence Mohr is a Fellow of the American College of Physicians and the American College of Chest Physicians. He is a member of several professional societies including: the American Federation for Medical Research; the Society for Risk Analysis; and the Wilderness Medical Society. Previously, he was on the Scientific Advisory Board for the Consortium in Environmental Risk Evaluation and the Savannah River Health Information System. He has authored or coauthored more than 60 articles, books, or technical publications. He received an A.B. degree in Chemistry as well as an M.D. degree, both from the University of North Carolina, Chapel Hill. Lawrence Mohr, Jr., is certified by the American Board of Internal Medicine.

Goetz K. Oertel's career in engineering, physics, chemistry, astronomy, and technical program management spans more than 40 years. He consults for industrial, academic, and governmental organizations in North and South America. As President and CEO of the Association of Universities for Research Astronomy, a nonprofit corporation, he engineered the initiation and completion of two 8-m aperture optical telescopes, and oversaw the Space Telescope Science Institute from before launch, through repair of the "Hubble flaw", to its successful operation. He initiated the conceptional phase of the Next Generation Space Telescope that will succeed Hubble as well as the Advanced Solar Telescope, and he oversaw the completion of ambitious ground-based astronomy facilities. He held technical and management positions in the U.S. Department of Energy, including Director of Defense Waste Management; Acting Manager of the Savannah River Operations Office; Deputy Manager of Albuquerque Operations Office; and Deputy Assistant Secretary for Safety, Health, and Quality Assurance. He had primary responsibility for the congressionally-mandated Defense Waste Management Plan, and for managing the related technology development, operations, and projects. He led the initiation of the Defense Waste Processing Facility, and saw it and the Waste Isolation Pilot Plant through technical, managerial, stakeholder, and political challenges. He was National Aeronautics and Space Administration Space Science Chief and Program Manager, and Aerospace Engineer at Langley. He was a Fellow in the White House with the President's Science Advisor and the Office of Management and Budget's Space and Energy branch. He chaired the Westinghouse West Valley Corporation Technical Advisory Group for high-level nuclear waste vitrification and management before, during, and after that project's successful vitrification campaign. He is a member of the American Physical Society, Sigma Xi, and other professional organizations. He is a Fellow of the American Association for the Advancement of Science. He is Chair or member of boards and committees of the National Research Council; George Mason University; the American Society of Mechanical Engineers; International University Exchange; and Westinghouse West Valley Corporation. He is a founding member of the Editorial Board for "Data Science", the new international on-line journal of Codata. He published numerous peer-reviewed papers and was awarded two patents. Trained as electrical engineer and physicist, he received a Vordiplom in Physics and Chemistry from the University of Kiel while on German industrial and governmental scholarships, and a Ph.D. in Physics from University of Maryland at College Park under a Fulbright scholarship.

Donald A. Orth is currently President of DO, Inc. He has over 50 years of experience in the fields of nuclear materials production and waste management, primarily at the U.S. Department of Energy (DOE), Savannah River Site (SRS) under the DuPont Atomic Energy Division and the Westinghouse Savannah River Company. His work there included development of processes for plutonium production and separation of nuclear materials. Related work and publications have been on solvent extraction processes, actinide chemistry, air cleaning, process hazards, nuclear criticality safety, and environmental effects of nuclear operations. Substantial work has been in the above areas for other DOE sites, the U.S. Nuclear Regulatory Commission, and as a consultant to the Advisory Committees on Reactor Safety and Nuclear Waste. Since retirement in 1992 he has: reviewed waste issues and research for DOE; served on the SRS Health Effects Subcommittee that reviews the Center for Disease Control and Prevention study of the SRS regional effects; spent five years on a University of Chicago review team for the electrometallurgical treatment of nuclear fields by Argonne National Laboratory. Under DuPont he was appointed Departmental Fellow, DuPont's highest corporate technical rank. In 1990 he received the Glenn T. Seaborg Actinide Separations Award. Upon his retirement, the Westinghouse Savannah River Company created the "Donald A. Orth Award of Technical Excellence" to be awarded annually. He is a member of the American Chemical Society and American Nuclear Society, and has served on the Executive Committee of the ACS Division of Nuclear Chemistry and Technology. Donald Orth received his B.S. in Chemistry and Ph.D. in Nuclear Chemistry from the University of California at Berkeley.

Francis J. Patti is currently an independent consultant with expertise in the fields of Civil Engineering and Nuclear Engineering. He was Chief Nuclear Engineer at Burns&Roe. He has been involved in nine major nuclear power projects, and has also worked on hot cell facilities; research reactors; power reactors; radioactive waste facilities; and decommissioning projects. His work covers the full range of architect-engineer activities and includes both analytical work and systems designs. His consulting assignments have included working with the Princeton Plasma Physics Laboratory on the Tokomak Fusion Test Reactor; Duquesne Light Company on Beaver Valley 2; Public Service of New Hampshire on Seabrook; Gulf States Utilities on the River Bend Nuclear Plant; Toledo Edison on the Davis-Besse Nuclear Power Station; Korea

Power Engineering Company; North Carolina Electric Membership Corporation on Catawba Nuclear Units 1 and 2; and the Brookhaven National Laboratory. He was project leader on engineering a major modification to the Brookhaven Medical Research Reactor—to enhance its capability for boron neutron capture therapy to treat certain types of brain cancer. His research reactor experience includes major participation in the design of the Argonne Advanced Research Reactor, the Army Pulsed Reactor Facility, and the NBS Research Reactor. While on these assignments, Francis Patti reviewed, evaluated, or was involved with: providing welding engineering support at construction sites; air modeling, permitting, and hazardous waste issues; waste water treatment systems for power plants; emergency plans; procedures; safety analysis; tritium systems upgrade engineering; architect-engineering capabilities; and the status of piping verification work. Francis Patti has served as Chair of the local sections of the American Nuclear Society and the American Society of Civil Engineers. He received Civil Engineer of the Year and Distinguished Service Awards from the local branch of the American Society of Civil Engineers. He is the author or coauthor of 15 technical papers. He holds a B.S. degree in Civil Engineering from Drexel University, Philadelphia, PA and an M.S. degree in Civil Engineering from the Massachusetts Institute of Technology, Cambridge, MA. He also has a diploma in Nuclear Science and Engineering from the International School of Nuclear Science and Engineering as well as M.S. and Professional Degrees in Nuclear Engineering from Columbia University, New York, NY. Francis Patti is a registered Professional Engineer in New York.

Joseph R. Peterson is Professor Emeritus of Chemistry at the University of Tennessee, Knoxville, TN. While at the university, he was also involved in the Transuranium Elements Research program at the Oak Ridge National Laboratory (ORNL). He directed the research and managed his undergraduate, graduate, and postdoctoral research associates from the University of Tennessee in the ORNL Transuranium Research Laboratory. His group determined, interpreted, and correlated the basic physical-chemical properties of the transuranium (and related 4f) elements and studied the similarities, and differences, of properties of these elements and their analogous lanthanide elements. A significant feature of his group's efforts was the application of several investigative techniques to the same transuranium element sample in order to characterize it more fully and understand its behavior more completely. His studies included evaluation of transuranium metals, oxides, halides, and other compounds, and their characterization by X-ray diffraction (angle- and energy-dispersive and as a function of temperature or pressure); various optical spectroscopic techniques such as absorption, phonon Raman, and luminescence; as well as magnetic susceptibility and solution calorimetry. These studies generated fundamental data that are now included in handbooks of chemical and physical data. Joseph Peterson's scientific career includes his extensive collaboration with foreign scientists and European laboratories. He has been a Visiting Scientist at the Nuclear Research Center in Karlsruhe, Germany; the European Institute for Transuranium Elements in Karlsruhe, Germany; and the Institute for Nuclear Chemistry at Johannes-Gutenberg-University in Mainz, Germany. Joseph Peterson has served as Treasurer, Chair, and now Councilor of the Division of Nuclear Chemistry and Technology of the American Chemical Society. He is also a lifetime member of the American Nuclear Society. He was awarded a NATO Postdoctoral Fellowship in Science in 1969. He is the author or co-author of over 200 publications. He received an A.B. degree in Chemistry from Swarthmore College, Swarthmore, PA and a Ph.D. in Chemistry from the University of California, Berkeley, CA.

Alfred Schneider is Professor Emeritus of Nuclear Engineering at the Georgia Institute of Technology. His interests are the nuclear fuel cycle, nuclear chemical engineering, and the management of radioactive wastes. Previously, he was a Visiting Professor of Nuclear Engineering and a Research Affiliate at the Massachusetts Institute of Technology (MIT). At MIT, he was responsible for organizing educational and research programs in nuclear chemical engineering and radioactive waste management. Prior to his association with MIT, he spent fifteen years at the Georgia Institute of Technology as a Professor of Nuclear Engineering. He was responsible for developing research programs and courses in nuclear engineering design and in the nuclear fuel cycle and energy conversion areas. Prior to his academic career, he was: Director of Nuclear Technology for Allied Chemical and Allied-General Nuclear Services; Manager of Materials and Processes for NUS; Manager of Nuclear Research and

Development at the Martin Marietta Corporation; and a Chemical Engineer/Chemist at Argonne National Laboratory, the Celanese Corporation, and the United States Testing Co. He has consulted for national laboratories, government organizations, and private industry in the areas of nuclear fuel cycle and isotopes separation, waste vitrification, and other forms of waste treatment. Alfred Schneider is a Fellow of the American Institute of Chemists. He was a member of the Executive Committee of the Fuel Cycle and Waste Management Division of the American Nuclear Society; a member of the Task Force on Radioactive Waste Management of the Advisory Board of the U.S. Secretary of Energy; Chairman of the Nuclear Engineering Division of the American Institute of Chemical Engineers; and a member of the Editorial Board of Annals of Nuclear Energy. He was a member of the National Research Council's committees on: Radioactive Waste Management; Environmental Management Technologies; Evaluation of the Science, Engineering, and Health Bases of the U.S. Department of Energy's Environmental Management Programs; Decontamination & Decommissioning (D&D) of Uranium Enrichment Facilities; and Development of Technologies for the D&D of Nuclear Facilities. He was awarded the U.S. Navy Antarctica Medal; the Robert E. Wilson Award of the American Institute of Chemical Engineers; and the Gano Dunn Medal from Cooper Union. He holds nine patents and is the co-inventor of an electrochemical method for the separation of uranium and plutonium. He has published his work in over 70 scientific and technical papers, reports, and conference transactions. Alfred Schneider received a B.S. in Chemical Engineering from Cooper Union in New York, and a Ph.D. in Chemical Engineering from Polytechnic University of New York.

Sorin R. Straja is currently Vice President for Science and Technology of the Institute for Regulatory Science. He has over 20 years of expertise in mathematical modeling and software development as applied in chemical engineering and risk assessment. Previously he served as Assistant Professor of Biostatistics with Temple University, Philadelphia; as Director of the Department of Occupational Health and Safety of Temple University, Philadelphia; and as a chemist with University of Maryland at Baltimore. Sorin Straja has extensive experience in the chemical industry where he worked as a senior R&D consultant with the Chemical and Biochemical Energetics Institute, and as a plant manager with Chemicals Enterprise Dudesti and Plastics Processing Bucharest from Romania. He was an Assistant/Adjunct Professor of Chemical Engineering with the Polytechnic Institute Bucharest. Sorin Straja is the author of two books and 44 scientific papers published in internationally recognized and peer-reviewed journals. He was an editor of *Environment International,* and currently is a contributing editor of *Technology.* Sorin Straja received a Certificate of Appreciation for Teaching from Temple University, the "Nicolae Teclu" Prize of the Romanian Academy, and a Certificate of Appreciation from U.S. Department of Agriculture for significant volunteer contributions. He is a Fellow of the Global Association of Risk Professionals, and a member of the American Chemical Society, American Institute of Chemical Engineers, Society for Risk Analysis, and New York Academy of Sciences. Sorin Straja holds a M.S. in Industrial Chemistry and a Ph.D. in Chemical Engineering both from Polytechnic Institute Bucharest.

Glenn W. Suter, II is currently Science Advisor at the National Center for Environmental Assessment of the U.S. Environmental Protection Agency (EPA) in Cincinnati, OH. Previous to his current position, he was at Oak Ridge National Laboratory, initially as Research Associate and gradually rising to Science Leader at the Environment Science Division of the Laboratory. His interest has focused on Ecotoxicology in general and Ecological Risk Assessment in particular. He is one of the developers of the most widely-used methodology for Ecological Risk Assessment. This method has been applied to the impact of pollutants on fish, contaminated soils, production of synthetic fuels, and various other ecosystems. Glenn Suter has lectured widely, both nationally and internationally on Ecological Risk Assessment. He is currently a member of the U.S. EPA's Risk Assessment Forum. He has been a member of numerous panels and has consulted with various governmental agencies and private organizations, including the Council of Environmental Quality. He was a member of the Scientific Review Panel for Savannah River Ecology Laboratory; the National Science Foundation Panel on Decision Making and Valuation for Environmental Policy; and the U.S. EPA Science Advisory Board and Conservation Foundation, Ecosystem Valuation Forum. In addition, he was a member of the International

Institute of Applied Systems Analysis Task Force on Risk and Policy Analysis and the Council on Environmental Quality. He was a member of the Board of Directors, for the Society for Environmental Toxicology and Chemistry. Glenn Suter is presently on the Editorial Board of *Environmental Health Perspectives* and *Human and Ecological Risk Assessment.* Previously, he was on the Editorial Board of *Handbook of Environmental Risk Assessment and Management* and *Environmental Toxicology and Chemistry.* Glenn Suter is the author of three books and is author and coauthor of over 200 publications. He received a B.S. degree in Biology from Virginia Polytechnic Institute and a Ph.D. in Ecology from the University of California, Davis.

Cheryl A. Trottier is currently Chief of the Radiation Protection, Environmental Risk, and Waste Management Branch of the Office of Nuclear Regulatory Research at the U.S. Nuclear Regulatory Commission. In that capacity, she is responsible for the management of research programs and the development of technical bases to support rulemaking. This includes the development of models for assessing the maximum doses likely to be received from lands and materials cleared from regulatory control; evaluating hydrologic model and parameter uncertainty; the development of realistic parameters for assessing sorption processes in geochemical models; and refining evaluations of radionuclide transport mechanism in the environment. In her 25 years of experience in the field of radiation protection, she has been involved in the management of environmental radiation protection monitoring programs and laboratory measurements, and the emergency preparedness coordination for an electric utility. She was also involved in the areas of materials use regulation oversight; development of regulations; and the development of guidance for use of byproduct and special nuclear materials. This included finalization of regulations and the development of regulations to certify the gaseous diffusion plants. Cheryl Trottier serves as one of the U.S. representatives to the Nuclear Energy Agency Committee on Radiation Protection and Public Health. She is a member of the American Nuclear Society, and serves as a member of the Committee on Site Clean-Up Restoration Standards. She received her B.A. degree in Biology from Rutgers University.

Charles O. Velzy is a consultant in the field of waste treatment and disposal. Previously, he held increasingly responsible positions with the environmental consulting engineering firm, Charles R. Velzy Associates, Inc., becoming President in 1976. In 1987, when Velzy Associates merged with Roy F. Weston, Inc., Charles Velzy became Vice President of Weston, a position which he held until retiring in 1992. He has over 35 years of experience as an environmental engineering consultant specializing in: the analysis of waste management problems; design of wastewater treatment and waste disposal systems; and design of new, retrofit of existing, testing, and permitting of waste combustion facilities. He has authored or co-authored over 80 publications—primarily in the field of solid waste management. He has served on the Science Advisory Board of the U.S. Environmental Protection Agency; as President of the American Society of Mechanical Engineers (ASME); Chair of the ASME Peer Review Committee; and as Treasurer of the American Academy of Environmental Engineers (AAEE). He has served on numerous committees of the ASME, the AAEE, the American National Standards Institute, and the American Society for Testing and Materials. He is a registered professional engineer in New York and eleven other states. Charles Velzy received B.S. degrees in Mechanical and Civil Engineering, and an M.S. in Sanitary Engineering from the University of Illinois at Urbana-Champaign.

Acronyms

ANL	Argonne National Laboratory
ARP	Actinide Removal Process
ASME	American Society of Mechanical Engineers
AST	Alpha Sorption Tank
BOBCalixC6	calix[4]arene-bis(tert-octylbenzo-crown-6)
CD-1	Critical Decision One
CF	concentration factor
CIF	Consolidated Incineration Facility
CINC	Costner Industries Nevada Corporation
Cs-7SB	1-(2,2,3,3, -tetrafluoropropoxy)-3-(4-sec-butylphenoxy)-2-propanol
CSDT	Cleaning Solution Dump Tank
CST	Crystalline Silicotitanate
CSSX	Caustic-Side Solvent Extraction
CUF	Cells Unit Filter
DF	decontamination factor
DNFSB	Defense Nuclear Facilities Safety Board
DOE	U.S. Department of Energy
DOE-SR	U.S. Department of Energy-Savannah River Operations Office
DSS	decontaminated salt solution
DWPF	Defense Waste Processing Facility
EP	Executive Panel
EPA	U.S. Environmental Protection Agency
EPC	Engineering, Procurement, and Construction
ESP	Extended Sludge Processing
ESS	extraction, scrubbing, and stripping
ETF	Effluent Treatment Facility
FFA	Federal Facilities Agreement
FFT	Filter Feed Tank
FRED	Filtration Research Engineering Demonstration
FY	Fiscal Year
HLW	High Level Waste
ICP-MS	Inductively Coupled Plasma-Mass Spectrometry
ITP	In-Tank Precipitation
LLW	low-level radioactive waste
LPPP	Low Point Pump Pit
MST	monosodium titanate

NAS	National Academy of Sciences
NRC	National Research Council
O/A	organic/aqueous flow ratios
ORNL	Oak Ridge National Laboratory
PRCEE	Peer Review Committee for Energy and Environment
PREF	Parallel Rheology Experimental Filter
RP	Review Panel
RSI	Institute for Regulatory Science
SCDHEC	South Carolina Department of Health and Environmental Control
SDF	Saltstone Disposal Facility
SPF	Saltstone Production Facility
SPP	Salt Processing Project
SRAT	Slurry Receipt Adjustment Tank
SROO	Savannah River Operations Office
SRS	Savannah River Site
SRTC	Savannah River Technology Center
SSFT	Salt Solution Feed Tank
SSRT	Sludge Slurry Receipt Tank
STP	Site Treatment Plan
STTP	Small Tank Tetraphenylborate Precipitation
SWPF	Salt Waste Processing Facility
TFA	Tanks Focus Area
TMP	transmembrane pressure
TOA	tri-*n*-octylamine
TWG	Technical Working Group
USNRC	U.S. Nuclear Regulatory Commission
WAC	Waste Acceptance Criteria
WSRC	Westinghouse Savannah River Company